101 Problems in Calculating Trigonometric Limits with Solutions

by

Richard Shedenhelm

INTRODUCTION

One of the early challenges for the beginning calculus student is computing trigonometric limits. I hope that working through the problems in this book will make such a student a master of the topic.

There are three fundamental math facts the student must automatize before starting on the problems:

$$1.\ \lim_{\theta \to 0} \frac{\sin(\theta)}{\theta} = 1 \qquad\qquad 2.\ \lim_{\theta \to 0} \sin(\theta) = 0 \qquad\qquad 3.\ \lim_{\theta \to 0} \cos(\theta) = 1$$

The chances are very good that the student enjoyed a thirty-minute lecture in class where the first fact was proved using the Sandwich Theorem (also known as the Squeeze Theorem). The student may safely forget the proof, but not the result, for that fact shows up in the vast majority of trigonometric limits. The two other facts are true because sine and cosine are continuous functions and $\sin(0) = 0$ and $\cos(0) = 1$.

The second set of math facts the student will need are some of the trigonometric identities:

$$\tan(\theta) = \frac{\sin(\theta)}{\cos(\theta)} \qquad\qquad\qquad \cot(\theta) = \frac{\cos(\theta)}{\sin(\theta)}$$

$$\sec(\theta) = \frac{1}{\cos(\theta)} \qquad\qquad \csc(\theta) = \frac{1}{\sin(\theta)} \qquad\qquad \cot(\theta) = \frac{1}{\tan(\theta)}$$

$$\sin{}^2(\theta) + \cos{}^2(\theta) = 1$$

The final mathematical prerequisite concerns fractions. There is a crucial fact about fractions that comes up in almost every trigonometric limit problem: We multiply fractions by multiplying the numerators and denominators. Hence, the order of the factors, say, in the numerator, can be whatever we please. For example, if we wish to multiply

$$\frac{2}{3} \cdot \frac{5}{7}$$

we could, if we desire, modify the fractions to look like

$$\frac{5}{3} \cdot \frac{2}{7}.$$

In either case, the multiplication would result in the answer

$$\frac{10}{21}.$$

Although this sort of manipulation of factor order is not usually important in ordinary arithmetic, the solution to a trigonometric limit often requires manipulating the order of the factors to fit certain arrangements that we can compute the limits of. For example, in the problem

$$\lim_{\theta \to 0} \frac{\sin(\theta)}{1} \cdot \frac{\cos(\theta)}{\theta}$$

it is helpful to rewrite the problem as

$$\lim_{\theta \to 0} \frac{\sin(\theta)}{\theta} \cdot \frac{\cos(\theta)}{1}$$

so that we may apply the limit laws and the basic trigonometric limits, viz.,

$$\lim_{\theta \to 0} \frac{\sin(\theta)}{\theta} \cdot \frac{\cos(\theta)}{1} = \lim_{\theta \to 0} \frac{\sin(\theta)}{\theta} \cdot \lim_{\theta \to 0} \cos(\theta) = (1)(1) = 1.$$

The reader will see this issue come up repeatedly in the solutions to the problems.

I have organized the problems into groups. Problems 1-11 I call the "Sine Group," problems 12-18 the "Cosine Group," problems 19-28 the "Cosine-Conjugate Group," problems 29-36 the "Tangent Group," problems 37-46 the "Trig-Identity Heavy Group," and problems 47-101 the "Miscellaneous Group." This grouping will help the student see some common characteristics among many of the problems. In the "Trig-Identity Heavy Group" there are some problems that use advanced trigonometric identities, such as the addition/subtraction formulas.

Occasionally in the solutions I use a designation such as "$\overset{\#1}{=}$" to point the reader back to an earlier solution.

Athens, Georgia, May 22, 2015.

Richard Shedenhelm

PROBLEMS

Find the following limits.

1. $\lim\limits_{\theta\to 0}\dfrac{\theta}{\sin(\theta)}$

2. $\lim\limits_{\theta\to 0}\dfrac{3\sin(\theta)}{\theta}\cdot$

3. $\lim\limits_{\theta\to 0}\dfrac{\sin(3\theta)}{\theta}$

4. $\lim\limits_{\theta\to 0}\dfrac{\sin(3\theta)}{\sin(4\theta)}$

5. $\lim\limits_{\theta\to 0^+}\dfrac{\theta}{\sin(\sqrt{\theta})}$

6. $\lim\limits_{\theta\to 0^+}\dfrac{\sin(\theta)}{\sqrt{\theta}}$

7. $\lim\limits_{\theta\to 0}\theta\csc(\theta)$

8. $\lim\limits_{\theta\to 0}\dfrac{\sin(-3\theta)}{4\theta}$

9. $\lim\limits_{\theta\to 0}\dfrac{\sin(a\theta)}{b\theta}$

10. $\lim\limits_{\theta\to 0}\dfrac{\sin(a\theta)}{\sin(b\theta)}$

11. $\lim\limits_{\theta\to 0}\dfrac{\sin^2(k\theta)}{\theta^2}$

12. $\lim\limits_{\theta\to 0}\sec(\theta)$

13. $\lim\limits_{\theta\to 0}3\cos(\theta)$

14. $\lim\limits_{\theta\to 0}\dfrac{1}{\cos(\theta)}$

15. $\lim\limits_{\theta\to 0}\dfrac{\cos(\theta)}{\theta}$

16. $\lim\limits_{\theta\to 0}\dfrac{\theta}{\cos(\theta)}$

17. $\lim\limits_{\theta\to 0}\dfrac{\cos(3\theta)}{\cos(4\theta)}$

18. $\lim\limits_{\theta\to 0}\dfrac{\cos(a\theta)}{\cos(b\theta)}$

19. $\lim\limits_{\theta\to 0}\dfrac{1-\cos(\theta)}{\theta}$

20. $\lim\limits_{\theta\to 0}\dfrac{\cos(\theta)-1}{\theta}$

21. $\lim\limits_{\theta\to 0}\dfrac{1-\cos(\theta)}{\theta^{\frac{2}{3}}}$

22. $\lim\limits_{\theta\to 0}\dfrac{2\cos(\theta)-2}{3\theta}$

23. $\lim\limits_{\theta\to 0}\dfrac{\theta^2}{1-\cos(\theta)}$

24. $\lim\limits_{\theta\to 0}\dfrac{1-\cos(5\theta)}{\cos(7\theta)-1}$

25. $\lim\limits_{\theta\to 0}\dfrac{1-2\theta^2-2\cos(\theta)+\cos^2(\theta)}{\theta^2}$

26. $\lim\limits_{\theta\to 0}\dfrac{1-\cos(\theta)}{\theta^2}$

27. $\lim\limits_{\theta\to 0}\dfrac{1-\cos(3\theta)}{\theta^2}$

28. $\lim\limits_{\theta\to 0}\dfrac{1-\cos(a\theta)}{b\theta}$

29. $\lim\limits_{\theta\to 0}\tan(\theta)$

30. $\lim\limits_{\theta\to 0}3\tan(\theta)$

31. $\lim\limits_{\theta\to 0}\dfrac{\tan(\theta)}{\theta}$

32. $\lim\limits_{\theta\to 0}\dfrac{\theta}{\tan(\theta)}$

33. $\lim\limits_{\theta\to 0}\dfrac{\tan(3\theta)}{\theta}$

34. $\lim\limits_{\theta\to 0}\theta\cot(\theta)$

35. $\lim\limits_{\theta\to 0}\dfrac{\tan(3\theta)}{\tan(4\theta)}$

36. $\lim\limits_{\theta\to 0}\dfrac{\tan(a\theta)}{\tan(b\theta)}$

37. $\lim\limits_{\theta\to 0}\dfrac{\theta}{\cos\left(\frac{\pi}{2}-\theta\right)}$

38. $\lim\limits_{\theta\to 0}\dfrac{\cos\left(\theta+\frac{\pi}{2}\right)}{\theta}$

39. $\lim\limits_{\theta\to 0}\dfrac{\cos\left(\frac{\pi}{2}-\theta\right)}{\theta}$

40. $\lim\limits_{\theta\to 0}\dfrac{\theta}{\sin\left(\theta+\frac{\pi}{2}\right)}$

41. $\lim\limits_{\theta\to 0}\dfrac{\sin\left(\frac{\pi}{2}-\theta\right)\tan(\pi+\theta)}{\theta}$

42. $\lim\limits_{\theta\to 0}\sin\left(\frac{\pi}{2}+\theta\right)\cos\left(\frac{\pi}{2}+\theta\right)$

43. $\lim\limits_{\theta\to 0}\dfrac{\cos\left(\frac{\pi}{2}-a\theta\right)}{b\theta}$

44. $\lim\limits_{\theta\to 0}\dfrac{\cos\left(\frac{\pi}{2}-a\theta\right)}{\cos\left(\frac{\pi}{2}-b\theta\right)}$

45. $\lim\limits_{\theta\to 0}\dfrac{\sin\left(\frac{\pi}{2}-a\theta\right)}{\sin\left(\frac{\pi}{2}-b\theta\right)}$

46. $\lim\limits_{\theta\to 0}[\sin(\pi-\theta)+\tan(\pi-\theta)]$

47. $\lim\limits_{\theta\to 0}\dfrac{\tan(a\theta)}{\sin(b\theta)}$

48. $\lim\limits_{\theta\to 0}\dfrac{\sin(\theta)}{1-\cos(\theta)}$

49. $\lim\limits_{\theta\to 0}\dfrac{5\theta+\sin(3\theta)}{\tan(4\theta)-7\theta\cos(6\theta)}$

50. $\lim\limits_{\theta\to 0}\dfrac{1-\cos^3(\theta)}{\sin^2(\theta)}$

51. $\lim\limits_{\theta\to 0}\dfrac{2+\sin(\theta)}{3+\theta}$

5

52. $\lim\limits_{\theta \to 0} \dfrac{\theta^2+1}{\theta+\cos(\theta)}$

53. $\lim\limits_{\theta \to 0} \dfrac{\theta+\tan(\theta)}{\sin(\theta)}$

54. $\lim\limits_{\theta \to 0} \dfrac{1-\cos(3\theta)}{\theta\sin(\theta)}$

55. $\lim\limits_{\theta \to 0} \dfrac{1-\cos(2\theta)}{\theta\sin(\theta)}$

56. $\lim\limits_{\theta \to 0} \dfrac{1-\cos(\theta)}{\theta\sin(\theta)}$

57. $\lim\limits_{\theta \to 0} \dfrac{\cos(\theta)}{\theta\csc(\theta)}$

58. $\lim\limits_{\theta \to 0} \sin(2\theta)\cot(\theta)$

59. $\lim\limits_{\theta \to 0} \dfrac{1-\cos(\theta)}{\sin(\theta)}$

60. $\lim\limits_{\theta \to 0} \dfrac{1-\cos(\theta)}{\tan(\theta)}$

61. $\lim\limits_{\theta \to 0} \dfrac{\sin(\theta)}{\tan(\theta)}$

62. $\lim\limits_{\theta \to 0} \dfrac{2\sin(\theta)-\sin(2\theta)}{\theta\cos(\theta)}$

63. $\lim\limits_{\theta \to 0} \dfrac{\tan(\theta)-\sin(\theta)}{\theta\cos(\theta)}$

64. $\lim\limits_{\theta \to 0} \dfrac{\csc(\theta)-\cot(\theta)}{\sin(\theta)}$

65. $\lim\limits_{\theta \to 0} \dfrac{2\theta+1-\cos(\theta)}{3\theta}$

66. $\lim\limits_{\theta \to 0} \dfrac{\sin^3(\theta)}{(2\theta)^3}$

67. $\lim\limits_{\theta \to 0} \dfrac{4\theta^2+3\theta\sin(\theta)}{\theta^2}$

68. $\lim\limits_{\theta \to 0} \dfrac{\sin[\cos(\theta)]}{\sec(\theta)}$

69. $\lim\limits_{\theta \to 0} \dfrac{\theta^2}{1-\cos^2(2\theta)}$

70. $\lim\limits_{\theta \to 0} \dfrac{\sec(6\theta)\tan(3\theta)}{\theta}$

71. $\lim\limits_{\theta \to 0} \theta^2\cot^2(4\theta)$

72. $\lim\limits_{\theta \to 0} \dfrac{\tan(\pi-\theta)-\theta}{\sin(\theta+\pi)}$

73. $\lim\limits_{\theta \to 0} \dfrac{\cos(a\theta)\tan(a\theta)}{b\theta}$

74. $\lim\limits_{\theta \to 0} \dfrac{\cos(a\theta)\tan(a\theta)}{\cos(b\theta)\tan(b\theta)}$

75. $\lim\limits_{\theta \to 0} \dfrac{\sin^2(\theta)+2\cos(\theta)-2}{\cos^2(\theta)-\sin(\theta)-1}$

76. $\lim\limits_{\theta \to 0} \dfrac{\sin(2\theta)-\tan(2\theta)}{\theta^2}$

77. $\lim\limits_{\theta \to 0} \dfrac{\sin(\theta)-2\theta}{\theta}$

78. $\lim\limits_{\theta \to 0} \dfrac{3-\csc(\theta)}{7-\cot(\theta)}$

79. $\lim\limits_{\theta \to 0} \dfrac{\theta\cos(\theta)-\sin(\theta)}{\theta}$

80. $\lim\limits_{\theta \to 0} \dfrac{\sin(2\theta)\tan(\theta)}{3\theta}$

81. $\lim\limits_{\theta \to 0} \dfrac{\sin(2\theta)+\tan(\theta)}{3\theta}$

82. $\lim\limits_{\theta \to 0} \dfrac{\tan(\theta)-\sin(\theta)}{\theta^2}$

83. $\lim\limits_{\theta \to 0} [\csc(\theta)-\cot(\theta)]$

84. $\lim\limits_{\theta \to 0} \left[\dfrac{1}{\theta^2}-\dfrac{1}{\theta^2\sec(\theta)}\right]$

85. $\lim\limits_{\theta \to 0} \dfrac{\sin(\theta)}{\theta+\theta^2}$

86. $\lim\limits_{\theta \to 0} \dfrac{\sec(\theta)-1}{\theta^2}$

87. $\lim\limits_{\theta \to 0} \dfrac{\cos(2\theta)-\cos(\theta)}{\sin^2(\theta)}$

88. $\lim\limits_{\theta \to 0} \dfrac{\cos(\theta)}{\csc(\theta)}$

89. $\lim\limits_{\theta \to 0} \dfrac{\theta^3}{\csc(\theta)+1}$

90. $\lim\limits_{\theta \to 0} 2\theta^2\sec^2(\theta)\cot^2(\theta)$

91. $\lim\limits_{\theta \to 0} \dfrac{\cot^4(\theta)\tan(\theta)+\sin^2(\theta)-\csc(\theta)+\sec(\theta)]}{\theta^{-3}}$

92. $\lim\limits_{\theta \to 0} \left[3\sec(\theta)-\dfrac{\theta^3\csc^3(\theta)}{\cos^3(\theta)}+\theta^2\csc(\theta)\right]$

93. $\lim\limits_{\theta \to 0} [\cos(\theta)-\sin^3(\theta)\csc^2(\theta)-\tan(\theta)]$

94. $\lim\limits_{\theta \to 0} [8\theta^2\csc^2(\theta)+\tan(\theta)\cos(\theta)]$

95. $\lim\limits_{\theta \to 0} 2\theta\cot(\theta)\sec(\theta)$

96. $\lim\limits_{\theta \to 0} \left[\dfrac{\cot(\theta)}{\csc(\theta)}+\sec(\theta)\right]$

97. $\lim\limits_{\theta \to 0} \dfrac{\sin(2\theta)\cos^3(\theta)}{\sin(\theta)}$

98. $\lim\limits_{\theta \to 0} [\cos^2(\theta)-\sec(\theta)\sin(\theta)]$

99. $\lim\limits_{\theta \to 0} \dfrac{\cot^2(\theta)+1}{\csc^2(\theta)}$

100. $\lim\limits_{\theta \to 0} \dfrac{\csc^2(\theta)}{\cot^2(\theta)+1}$

101. $\lim\limits_{\theta \to 0} \dfrac{\sec^2(\theta)}{\tan^2(\theta)+1}$

ANSWERS

1. 1

2. 3

3. 3

4. $\frac{3}{4}$

5. 0

6. 0

7. 1

8. $\frac{-3}{4}$

9. $\frac{a}{b}$

10. $\frac{a}{b}$

11. k^2

12. 1

13. 3

14. 1

15. Does not exist

16. 0

17. 1

18. 1

19. 0

20. 0

21. 0

22. 0

23. 2

24. $-\frac{25}{49}$

25. -2

26. $\frac{1}{2}$

27. $\frac{9}{2}$

28. 0

29. 0

30. 0

31. 1

32. 1

33. 3

34. 1

35. $\frac{3}{4}$

36. $\frac{a}{b}$

37. 1

38. -1

39. 1

40. 1

41. 1

42. 0

43. $\frac{a}{b}$

44. $\frac{a}{b}$

45. 1

46. 0

47. $\frac{a}{b}$

48. Does not exist

49. $\frac{8}{-3}$

50. $\frac{3}{2}$

51. $\frac{2}{3}$

52. 1

53. 2

54. $\frac{9}{2}$

55. 2

56. $\frac{1}{2}$

57. 1

58. 2

59. 0

60. 0

61. 1

62. 0

63. 0

64. $\frac{1}{2}$

65. $\frac{2}{3}$

66. $\frac{1}{8}$

67. 7

68. sin(1)

69. $\frac{1}{4}$

70. 3

71. $\frac{1}{16}$

72. 2

73. $\frac{a}{b}$

74. $\frac{a}{b}$

75. 0

76. 0

77. −1

78. 1

79. 0

80. 0

81. 1

82. 0

83. 0

84. $\frac{1}{2}$

85. 1

86. $\frac{1}{2}$

87. $\frac{-3}{2}$

88. 0

89. 0

90. 2

91. 1

92. 3

93. 1

94. 8

95. 2

96. 2

97. 2

98. 1

99. 1

100. 1

101. 1

SOLUTIONS

Sine Group:

1. $\displaystyle\lim_{\theta\to 0}\frac{\theta}{\sin(\theta)} = \lim_{\theta\to 0}\frac{1}{\left(\frac{\sin(\theta)}{\theta}\right)} = \frac{\displaystyle\lim_{\theta\to 0} 1}{\displaystyle\lim_{\theta\to 0}\left(\frac{\sin(\theta)}{\theta}\right)} = \frac{1}{1} = 1.$

2. $\displaystyle\lim_{\theta\to 0}\frac{3\sin(\theta)}{\theta} = \lim_{\theta\to 0} 3\cdot\frac{\sin(\theta)}{\theta} = \lim_{\theta\to 0} 3\cdot\lim_{\theta\to 0}\frac{\sin(\theta)}{\theta} = 3\cdot 1 = 3.$

3. $\displaystyle\lim_{\theta\to 0}\frac{\sin(3\theta)}{\theta} = \lim_{\theta\to 0}\frac{\sin(3\theta)}{\theta}\cdot\frac{3}{3} = \lim_{\theta\to 0}\frac{\sin(3\theta)}{3\theta}\cdot\frac{3}{1} = \lim_{\theta\to 0}\frac{\sin(3\theta)}{3\theta}\cdot\lim_{\theta\to 0} 3 = 1\cdot 3 = 3.$

Remark: $3\theta\to 0$ as $\theta\to 0$. That is why we can say $\displaystyle\lim_{\theta\to 0}\frac{\sin(3\theta)}{3\theta} = 1$. Similar cases will arise in some of the following solutions.

4. $\displaystyle\lim_{\theta\to 0}\frac{\sin(3\theta)}{\sin(4\theta)} = \lim_{\theta\to 0}\frac{\sin(3\theta)}{1}\cdot\frac{1}{\sin(4\theta)} = \lim_{\theta\to 0}\frac{\sin(3\theta)}{1}\cdot\frac{3\theta}{3\theta}\cdot\frac{1}{\sin(4\theta)}\cdot\frac{4\theta}{4\theta} =$

   $\displaystyle = \lim_{\theta\to 0}\frac{\sin(3\theta)}{3\theta}\cdot\frac{3\theta}{1}\cdot\frac{4\theta}{\sin(4\theta)}\cdot\frac{1}{4\theta} = \lim_{\theta\to 0}\frac{\sin(3\theta)}{3\theta}\cdot\frac{4\theta}{\sin(4\theta)}\cdot\frac{3\theta}{4\theta} = \lim_{\theta\to 0}\frac{\sin(3\theta)}{3\theta}\cdot\frac{4\theta}{\sin(4\theta)}\cdot\frac{3}{4} =$

   $\displaystyle = \lim_{\theta\to 0}\frac{\sin(3\theta)}{3\theta}\cdot\lim_{\theta\to 0}\frac{4\theta}{\sin(4\theta)}\cdot\lim_{\theta\to 0}\frac{3}{4} \overset{\#1}{=} 1\cdot 1\cdot\frac{3}{4} = \frac{3}{4}.$

5. $\displaystyle\lim_{\theta\to 0^+}\frac{\theta}{\sin(\sqrt{\theta})} = \lim_{\theta\to 0^+}\frac{\sqrt{\theta}}{\sin(\sqrt{\theta})}\cdot\frac{\sqrt{\theta}}{1} = \lim_{\theta\to 0^+}\frac{\sqrt{\theta}}{\sin(\sqrt{\theta})}\cdot\lim_{\theta\to 0^+}\sqrt{\theta} = \lim_{\theta\to 0^+}\frac{\sqrt{\theta}}{\sin(\sqrt{\theta})}\cdot\sqrt{\lim_{\theta\to 0^+}\theta} \overset{\#1}{=}$

   $\overset{\#1}{=} 1\cdot\sqrt{0} = 1\cdot 0 = 0.$

6. $\displaystyle\lim_{\theta\to 0^+}\frac{\sin(\theta)}{\sqrt{\theta}} = \lim_{\theta\to 0^+}\frac{\sin(\theta)}{\sqrt{\theta}}\cdot\frac{\sqrt{\theta}}{\sqrt{\theta}} = \lim_{\theta\to 0^+}\frac{\sin(\theta)}{\theta}\cdot\frac{\sqrt{\theta}}{1} = \lim_{\theta\to 0^+}\frac{\sin(\theta)}{\theta}\cdot\lim_{\theta\to 0^+}\sqrt{\theta} =$

   $\displaystyle = \lim_{\theta\to 0^+}\frac{\sin(\theta)}{\theta}\cdot\sqrt{\lim_{\theta\to 0^+}\theta} = 1\cdot\sqrt{0} = 1\cdot 0 = 0.$

9

7. $\displaystyle\lim_{\theta\to 0}\theta\csc(\theta) = \lim_{\theta\to 0}\theta\cdot\frac{1}{\sin(\theta)} = \lim_{\theta\to 0}\frac{\theta}{\sin(\theta)} \overset{\#1}{=} 1.$

8. $\displaystyle\lim_{\theta\to 0}\frac{\sin(-3\theta)}{4\theta} = \lim_{\theta\to 0}\frac{-\sin(3\theta)}{4\theta} = \lim_{\theta\to 0}\frac{-1}{4}\cdot\frac{\sin(3\theta)}{\theta} = -\frac{1}{4}\lim_{\theta\to 0}\frac{\sin(3\theta)}{\theta} =$

$\displaystyle = -\frac{1}{4}\lim_{\theta\to 0}\frac{\sin(3\theta)}{\theta}\cdot\frac{3}{3} = -\frac{3}{4}\lim_{\theta\to 0}\frac{\sin(3\theta)}{3\theta} = -\frac{3}{4}(1) = -\frac{3}{4}.$

Alternate Solution:

$\displaystyle\lim_{\theta\to 0}\frac{\sin(-3\theta)}{4\theta} = \frac{1}{4}\lim_{\theta\to 0}\frac{\sin(-3\theta)}{\theta} = \frac{1}{4}\lim_{\theta\to 0}\frac{\sin(-3\theta)}{\theta}\cdot\frac{-3}{-3} = \frac{-3}{4}\lim_{\theta\to 0}\frac{\sin(-3\theta)}{-3\theta} =$

$\displaystyle = \frac{-3}{4}\cdot(1) = \frac{-3}{4}.$

Let a and b be nonzero numbers.

9. $\displaystyle\lim_{\theta\to 0}\frac{\sin(a\theta)}{b\theta} = \lim_{\theta\to 0}\frac{1}{b}\cdot\frac{\sin(\theta)}{\theta} = \frac{1}{b}\lim_{\theta\to 0}\frac{\sin(a\theta)}{\theta} = \frac{1}{b}\lim_{\theta\to 0}\frac{\sin(a\theta)}{\theta}\cdot\frac{a}{a} = \frac{a}{b}\lim_{\theta\to 0}\frac{\sin(a\theta)}{a\theta}.$

$\displaystyle = \frac{a}{b}\cdot 1 = \frac{a}{b}.$

Let a and b be nonzero numbers.

10. $\displaystyle\lim_{\theta\to 0}\frac{\sin(a\theta)}{\sin(b\theta)} = \lim_{\theta\to 0}\frac{\sin(a\theta)}{1}\cdot\frac{1}{\sin(b\theta)} = \lim_{\theta\to 0}\frac{\sin(a\theta)}{1}\cdot\frac{1}{\sin(b\theta)}\cdot\frac{ab\theta}{ab\theta} =$

$\displaystyle = \frac{a}{b}\lim_{\theta\to 0}\frac{\sin(a\theta)}{a\theta}\cdot\frac{b\theta}{\sin(b\theta)} = \frac{a}{b}\lim_{\theta\to 0}\frac{\sin(a\theta)}{a\theta}\cdot\lim_{\theta\to 0}\frac{b\theta}{\sin(b\theta)} \overset{\#1}{=} \frac{a}{b}\cdot 1\cdot 1 = \frac{a}{b}.$

Let k be a nonzero number.

11. $\displaystyle\lim_{\theta\to 0}\frac{\sin^2(k\theta)}{\theta^2} = \lim_{\theta\to 0}\frac{\sin(k\theta)}{\theta}\cdot\frac{\sin(k\theta)}{\theta} = \lim_{\theta\to 0}\frac{\sin(k\theta)}{\theta}\cdot\frac{\sin(k\theta)}{\theta}\cdot\frac{k^2}{k^2} =$

$\displaystyle = \lim_{\theta\to 0}\frac{\sin(k\theta)}{k\theta}\cdot\frac{\sin(k\theta)}{k\theta}\cdot\frac{k^2}{1} = \lim_{\theta\to 0}\frac{\sin(k\theta)}{k\theta}\cdot\lim_{\theta\to 0}\frac{\sin(k\theta)}{k\theta}\cdot\lim_{\theta\to 0}\frac{k^2}{1} =$

$\displaystyle = \lim_{\theta\to 0}\frac{\sin(k\theta)}{k\theta}\cdot\lim_{\theta\to 0}\frac{\sin(k\theta)}{k\theta}\cdot k^2 = 1\cdot 1\cdot k^2 = k^2.$

Cosine Group:

12. $\displaystyle\lim_{\theta\to0}\sec(\theta)=\lim_{\theta\to0}\frac{1}{\cos(\theta)}=\frac{\displaystyle\lim_{\theta\to0}1}{\displaystyle\lim_{\theta\to0}\cos(\theta)}=\frac{1}{\cos(0)}=\frac{1}{1}=1.$

13. $\displaystyle\lim_{\theta\to0}3\cos(\theta)=\lim_{\theta\to0}3\cdot\lim_{\theta\to0}\cos(\theta)=3\cos(0)=3\cdot1=3.$

14. $\displaystyle\lim_{\theta\to0}\frac{1}{\cos(\theta)}=\frac{\displaystyle\lim_{\theta\to0}1}{\displaystyle\lim_{\theta\to0}\cos(\theta)}=\frac{1}{\cos(0)}=\frac{1}{1}=1.$

15. $\displaystyle\lim_{\theta\to0}\frac{\cos(\theta)}{\theta}=$ Does not exist.

Remark: Note that the numerator is approaching the fixed number 1 while the denominator is approaching 0. Furthermore, $\displaystyle\lim_{\theta\to0^+}\frac{\cos(\theta)}{\theta}=+\infty$ while $\displaystyle\lim_{\theta\to0^-}\frac{\cos(\theta)}{\theta}=-\infty.$

16. $\displaystyle\lim_{\theta\to0}\frac{\theta}{\cos(\theta)}=\frac{\displaystyle\lim_{\theta\to0}\theta}{\displaystyle\lim_{\theta\to0}\cos(\theta)}=\frac{0}{\cos(0)}=\frac{0}{1}=0.$

17. $\displaystyle\lim_{\theta\to0}\frac{\cos(3\theta)}{\cos(4\theta)}=\frac{\displaystyle\lim_{\theta\to0}\cos(3\theta)}{\displaystyle\lim_{\theta\to0}\cos(4\theta)}=\frac{\cos(0)}{\cos(0)}=\frac{1}{1}=1.$

18. $\displaystyle\lim_{\theta\to0}\frac{\cos(a\theta)}{\cos(b\theta)}=\frac{\displaystyle\lim_{\theta\to0}\cos(a\theta)}{\displaystyle\lim_{\theta\to0}\cos(b\theta)}=\frac{\cos(0)}{\cos(0)}=\frac{1}{1}=1.$

Cosine-Conjugate Group:

19. $\lim\limits_{\theta\to 0}\dfrac{1-\cos(\theta)}{\theta} = \lim\limits_{\theta\to 0}\dfrac{1-\cos(\theta)}{\theta}\cdot\dfrac{1+\cos(\theta)}{1+\cos(\theta)} = \lim\limits_{\theta\to 0}\dfrac{1-\cos^2(\theta)}{\theta[1+\cos(\theta)]} = \lim\limits_{\theta\to 0}\dfrac{\sin^2(\theta)}{\theta[1+\cos(\theta)]} =$

$= \lim\limits_{\theta\to 0}\dfrac{\sin(\theta)}{\theta}\cdot\dfrac{\sin(\theta)}{1}\cdot\dfrac{1}{1+\cos(\theta)} = \lim\limits_{\theta\to 0}\dfrac{\sin(\theta)}{\theta}\cdot\dfrac{\sin(\theta)}{1}\cdot\dfrac{\theta}{\theta}\cdot\dfrac{1}{1+\cos(\theta)} =$

$= \lim\limits_{\theta\to 0}\dfrac{\sin(\theta)}{\theta}\cdot\dfrac{\sin(\theta)}{\theta}\cdot\dfrac{\theta}{1}\cdot\dfrac{1}{1+\cos(\theta)} = \lim\limits_{\theta\to 0}\dfrac{\sin(\theta)}{\theta}\cdot\lim\limits_{\theta\to 0}\dfrac{\sin(\theta)}{\theta}\cdot\lim\limits_{\theta\to 0}\dfrac{\theta}{1}\cdot\lim\limits_{\theta\to 0}\dfrac{1}{1+\cos(\theta)} =$

$= \lim\limits_{\theta\to 0}\dfrac{\sin(\theta)}{\theta}\cdot\lim\limits_{\theta\to 0}\dfrac{\sin(\theta)}{\theta}\cdot\lim\limits_{\theta\to 0}\dfrac{\theta}{1}\cdot\dfrac{\lim\limits_{\theta\to 0}1}{\lim\limits_{\theta\to 0}[1+\cos(\theta)]} = 1\cdot 1\cdot 0\cdot\dfrac{1}{1+\cos(0)} = \dfrac{0}{1+1} = 0.$

20. $\lim\limits_{\theta\to 0}\dfrac{\cos(\theta)-1}{\theta} = \lim\limits_{\theta\to 0}\dfrac{\cos(\theta)-1}{\theta}\cdot\dfrac{\cos(\theta)+1}{\cos(\theta)+1} = \lim\limits_{\theta\to 0}\dfrac{\cos^2(\theta)-1}{\theta[\cos(\theta)+1]} = \lim\limits_{\theta\to 0}\dfrac{-[1-\cos^2(\theta)]}{\theta[\cos(\theta)+1]} =$

$= \lim\limits_{\theta\to 0}\dfrac{-\sin^2(\theta)}{\theta[\cos(\theta)+1]} = -\lim\limits_{\theta\to 0}\dfrac{\sin(\theta)}{\theta}\cdot\dfrac{\sin(\theta)}{1}\cdot\dfrac{1}{\cos(\theta)+1} =$

$= -\lim\limits_{\theta\to 0}\dfrac{\sin(\theta)}{\theta}\cdot\dfrac{\sin(\theta)}{1}\cdot\dfrac{\theta}{\theta}\cdot\dfrac{1}{\cos(\theta)+1} = \lim\limits_{\theta\to 0}\dfrac{\sin(\theta)}{\theta}\cdot\dfrac{\sin(\theta)}{\theta}\cdot\dfrac{\theta}{1}\cdot\dfrac{1}{\cos(\theta)+1} =$

$= \lim\limits_{\theta\to 0}\dfrac{\sin(\theta)}{\theta}\cdot\lim\limits_{\theta\to 0}\dfrac{\sin(\theta)}{\theta}\cdot\lim\limits_{\theta\to 0}\dfrac{\theta}{1}\cdot\lim\limits_{\theta\to 0}\dfrac{1}{\cos(\theta)+1} = 1\cdot 1\cdot 0\cdot\dfrac{\lim\limits_{\theta\to 0}1}{\lim\limits_{\theta\to 0}[\cos(\theta)+1]} =$

$= 1\cdot 1\cdot 0\cdot\dfrac{1}{\cos(0)+1} = \dfrac{0}{1+1} = 0.$

21. $\lim\limits_{\theta\to 0}\dfrac{1-\cos(\theta)}{\theta^{\frac{2}{3}}} = \lim\limits_{\theta\to 0}\dfrac{1-\cos(\theta)}{\theta^{\frac{2}{3}}}\cdot\dfrac{\theta^{\frac{1}{3}}}{\theta^{\frac{1}{3}}} = \lim\limits_{\theta\to 0}\dfrac{1-\cos(\theta)}{\theta}\cdot\theta^{\frac{1}{3}} = \lim\limits_{\theta\to 0}\dfrac{1-\cos(\theta)}{\theta}\cdot\lim\limits_{\theta\to 0}\theta^{\frac{1}{3}}\overset{\#19}{\triangleq}$

$\overset{\#19}{\triangleq} 0\cdot 0 = 0.$

22. $\lim\limits_{\theta\to 0}\dfrac{2\cos(\theta)-2}{3\theta} = \lim\limits_{\theta\to 0}\dfrac{2(\cos(\theta)-1)}{3\theta} = \lim\limits_{\theta\to 0}\dfrac{2}{3}\cdot\dfrac{\cos(\theta)-1}{\theta} = \dfrac{2}{3}\lim\limits_{\theta\to 0}\dfrac{\cos(\theta)-1}{\theta}\overset{\#20}{\triangleq}\dfrac{2}{3}\cdot 0 = 0.$

23. $$\lim_{\theta\to 0}\frac{\theta^2}{1-\cos(\theta)}=\lim_{\theta\to 0}\frac{\theta}{1}\cdot\frac{\theta}{1-\cos(\theta)}=\lim_{\theta\to 0}\frac{\theta}{1}\cdot\frac{\theta}{1-\cos(\theta)}\cdot\frac{1+\cos(\theta)}{1+\cos(\theta)}=$$

$$=\lim_{\theta\to 0}\frac{\theta}{1}\cdot\frac{\theta[1+\cos(\theta)]}{1-\cos^2\theta}=\lim_{\theta\to 0}\frac{\theta}{1}\cdot\frac{\theta[1+\cos(\theta)]}{\sin^2\theta}=\lim_{\theta\to 0}\frac{\theta}{\sin(\theta)}\cdot\frac{\theta}{\sin(\theta)}\cdot\frac{[1+\cos(\theta)]}{1}=$$

$$=\lim_{\theta\to 0}\frac{\theta}{\sin(\theta)}\cdot\lim_{\theta\to 0}\frac{\theta}{\sin(\theta)}\cdot\lim_{\theta\to 0}\frac{[1+\cos(\theta)]}{1}=\lim_{\theta\to 0}\frac{\theta}{\sin(\theta)}\cdot\lim_{\theta\to 0}\frac{\theta}{\sin(\theta)}\cdot\frac{\lim\limits_{\theta\to 0}[1+\cos(\theta)]}{\lim\limits_{\theta\to 0}1}\overset{\#1}{=}$$

$$\overset{\#1}{=}1\cdot 1\cdot\frac{1+\cos(0)}{1}=\frac{1+1}{1}=2.$$

24. $$\lim_{\theta\to 0}\frac{1-\cos(5\theta)}{\cos(7\theta)-1}=\lim_{\theta\to 0}\frac{1-\cos(5\theta)}{-[1-\cos(7\theta)]}=-\lim_{\theta\to 0}\frac{1-\cos(5\theta)}{1-\cos(7\theta)}\cdot\frac{1+\cos(5\theta)}{1+\cos(5\theta)}\cdot\frac{1+\cos(7\theta)}{1+\cos(7\theta)}=$$

$$=-\lim_{\theta\to 0}\frac{1-\cos^2(5\theta)}{1-\cos^2(7\theta)}\cdot\frac{1+\cos(7\theta)}{1+\cos(5\theta)}=-\lim_{\theta\to 0}\frac{\sin^2(5\theta)}{\sin^2(7\theta)}\cdot\frac{1+\cos(7\theta)}{1+\cos(5\theta)}=$$

$$=-\lim_{\theta\to 0}\frac{\sin^2(5\theta)}{\sin^2(7\theta)}\cdot\lim_{\theta\to 0}\frac{1+\cos(7\theta)}{1+\cos(5\theta)}=-\lim_{\theta\to 0}\frac{\sin^2(5\theta)}{\sin^2(7\theta)}\cdot\frac{\lim\limits_{\theta\to 0}[1+\cos(7\theta)]}{\lim\limits_{\theta\to 0}[1+\cos(5\theta)]}=$$

$$-\lim_{\theta\to 0}\frac{\sin^2(5\theta)}{\sin^2(7\theta)}\cdot\frac{1+\cos(0)}{1+\cos(0)}=-\lim_{\theta\to 0}\frac{\sin^2(5\theta)}{\sin^2(7\theta)}\cdot\frac{1+1}{1+1}=-\lim_{\theta\to 0}\frac{\sin^2(5\theta)}{\sin^2(7\theta)}\cdot\frac{2}{2}=$$

$$=-\lim_{\theta\to 0}\frac{\sin^2(5\theta)}{\sin^2(7\theta)}=-\lim_{\theta\to 0}\frac{\sin(5\theta)}{\sin(7\theta)}\cdot\frac{\sin(5\theta)}{\sin(7\theta)}=$$

$$=-\lim_{\theta\to 0}\frac{\sin(5\theta)}{1}\cdot\frac{\sin(5\theta)}{1}\cdot\frac{1}{\sin(7\theta)}\cdot\frac{1}{\sin(7\theta)}\cdot\frac{25\theta^2}{25\theta^2}\cdot\frac{49\theta^2}{49\theta^2}=$$

$$=-\lim_{\theta\to 0}\frac{\sin(5\theta)}{5\theta}\cdot\frac{\sin(5\theta)}{5\theta}\cdot\frac{7\theta}{\sin(7\theta)}\cdot\frac{7\theta}{\sin(7\theta)}\cdot\frac{25\theta^2}{49\theta^2}=$$

$$=-\lim_{\theta\to 0}\frac{\sin(5\theta)}{5\theta}\cdot\lim_{\theta\to 0}\frac{\sin(5\theta)}{5\theta}\cdot\lim_{\theta\to 0}\frac{7\theta}{\sin(7\theta)}\cdot\lim_{\theta\to 0}\frac{7\theta}{\sin(7\theta)}\cdot\lim_{\theta\to 0}\frac{25\theta^2}{49\theta^2}\overset{\#1}{=}$$

$$\overset{\#1}{=}-\left(1\cdot 1\cdot 1\cdot 1\cdot\frac{25}{49}\right)=-\frac{25}{49}$$

25. $$\lim_{\theta\to 0}\frac{1-2\theta^2-2\cos(\theta)+\cos^2(\theta)}{\theta^2}=\lim_{\theta\to 0}\frac{1-2\theta^2-2\cos(\theta)+1-\sin^2(\theta)}{\theta^2}=$$

$$=\lim_{\theta\to 0}\frac{2-2\theta^2-2\cos(\theta)-\sin^2(\theta)}{\theta^2}=\lim_{\theta\to 0}\left(\frac{2-2\cos(\theta)}{\theta^2}-\frac{2\theta^2+\sin^2(\theta)}{\theta^2}\right)=$$

$$=\lim_{\theta\to 0}\frac{2[1-\cos(\theta)]}{\theta^2}-\lim_{\theta\to 0}\frac{2\theta^2+\sin^2(\theta)}{\theta^2}=2\lim_{\theta\to 0}\frac{1-\cos(\theta)}{\theta^2}-\lim_{\theta\to 0}\frac{2\theta^2+\sin^2(\theta)}{\theta^2}=$$

$$= 2\lim_{\theta \to 0} \frac{1 - \cos(\theta)}{\theta^2} \cdot \frac{1 + \cos(\theta)}{1 + \cos(\theta)} - \lim_{\theta \to 0}\left(2 + \frac{\sin^2(\theta)}{\theta^2}\right) =$$

$$= 2\lim_{\theta \to 0} \frac{1 - \cos^2(\theta)}{\theta^2(1 + \cos(\theta))} - \lim_{\theta \to 0} 2 - \lim_{\theta \to 0} \frac{\sin^2(\theta)}{\theta^2} = 2\lim_{\theta \to 0} \frac{\sin^2(\theta)}{\theta^2(1 + \cos(\theta))} - 2 - 1 =$$

$$= 2\lim_{\theta \to 0} \frac{\sin^2(\theta)}{\theta^2} \cdot \frac{1}{1 + \cos(\theta)} - 3 = 2\lim_{\theta \to 0} \frac{\sin^2(\theta)}{\theta^2} \cdot \lim_{\theta \to 0} \frac{1}{1 + \cos(\theta)} - 3 =$$

$$= 2\lim_{\theta \to 0} \frac{\sin(\theta)}{\theta} \cdot \lim_{\theta \to 0} \frac{\sin(\theta)}{\theta} \cdot \frac{\lim_{\theta \to 0} 1}{\lim_{\theta \to 0}[1 + \cos(\theta)]} - 3 = 2\left(1 \cdot 1 \cdot \frac{1}{1 + \cos(0)}\right) - 3 =$$

$$= 2\left(\frac{1}{1 + 1}\right) - 3 = 2\left(\frac{1}{2}\right) - 3 = 1 - 3 = -2.$$

26. $$\lim_{\theta \to 0} \frac{1 - \cos(\theta)}{\theta^2} = \lim_{\theta \to 0} \frac{1 - \cos(\theta)}{\theta^2} \cdot \frac{1 + \cos(\theta)}{1 + \cos(\theta)} = \lim_{\theta \to 0} \frac{1 - \cos^2(\theta)}{\theta^2[1 + \cos(\theta)]} =$$

$$= \lim_{\theta \to 0} \frac{\sin^2(\theta)}{\theta^2(1 + \cos(\theta))} = \lim_{\theta \to 0} \frac{\sin(\theta)}{\theta} \cdot \frac{\sin(\theta)}{\theta} \cdot \frac{1}{1 + \cos(\theta)} =$$

$$= \lim_{\theta \to 0} \frac{\sin(\theta)}{\theta} \cdot \lim_{\theta \to 0} \frac{\sin(\theta)}{\theta} \cdot \lim_{\theta \to 0} \frac{1}{1 + \cos(\theta)} = \lim_{\theta \to 0} \frac{\sin(\theta)}{\theta} \cdot \lim_{\theta \to 0} \frac{\sin(\theta)}{\theta} \cdot \frac{\lim_{\theta \to 0} 1}{\lim_{\theta \to 0}[1 + \cos(\theta)]} =$$

$$= 1 \cdot 1 \cdot \frac{1}{1 + \cos(0)} = \frac{1}{1 + 1} = \frac{1}{2}.$$

27. $$\lim_{\theta \to 0} \frac{1 - \cos(3\theta)}{\theta^2} = \lim_{\theta \to 0} \frac{1 - \cos(3\theta)}{\theta^2} \cdot \frac{1 + \cos(3\theta)}{1 + \cos(3\theta)} = \lim_{\theta \to 0} \frac{1 - \cos^2(3\theta)}{\theta^2(1 + \cos(3\theta))} =$$

$$= \lim_{\theta \to 0} \frac{\sin^2(3\theta)}{\theta^2[1 + \cos(3\theta)]} = \lim_{\theta \to 0} \frac{\sin(3\theta)}{\theta} \cdot \frac{\sin(3\theta)}{\theta} \cdot \frac{1}{1 + \cos(3\theta)} =$$

$$= \lim_{\theta \to 0} \frac{\sin(3\theta)}{\theta} \cdot \frac{3}{3} \cdot \frac{\sin(3\theta)}{\theta} \cdot \frac{3}{3} \cdot \frac{1}{1 + \cos(3\theta)} = \lim_{\theta \to 0} \frac{\sin(3\theta)}{3\theta} \cdot \frac{\sin(3\theta)}{3\theta} \cdot \frac{9}{1 + \cos(3\theta)} =$$

$$= \lim_{\theta \to 0} \frac{\sin(3\theta)}{3\theta} \cdot \lim_{\theta \to 0} \frac{\sin(3\theta)}{3\theta} \cdot \lim_{\theta \to 0} \frac{9}{1 + \cos(3\theta)} =$$

$$= \lim_{\theta \to 0} \frac{\sin(3\theta)}{3\theta} \cdot \lim_{\theta \to 0} \frac{\sin(3\theta)}{3\theta} \cdot \frac{\lim_{\theta \to 0} 9}{\lim_{\theta \to 0}[1 + \cos(3\theta)]} = 1 \cdot 1 \cdot \frac{9}{1 + \cos(0)} = \frac{9}{1 + 1} = \frac{9}{2}.$$

Let a and b be nonzero numbers.

28. $\lim\limits_{\theta \to 0} \dfrac{1 - \cos(a\theta)}{b\theta} = \dfrac{1}{b}\lim\limits_{\theta \to 0} \dfrac{1 - \cos(a\theta)}{\theta} = \dfrac{1}{b}\lim\limits_{\theta \to 0} \dfrac{1 - \cos(a\theta)}{\theta} \cdot \dfrac{1 + \cos(a\theta)}{1 + \cos(a\theta)} =$

$= \dfrac{1}{b}\lim\limits_{\theta \to 0} \dfrac{1 - \cos^2(a\theta)}{\theta[1 + \cos(a\theta)]} = \dfrac{1}{b}\lim\limits_{\theta \to 0} \dfrac{\sin^2(a\theta)}{\theta[1 + \cos(a\theta)]} = \dfrac{1}{b}\lim\limits_{\theta \to 0} \dfrac{\sin(a\theta)}{\theta} \cdot \dfrac{\sin(a\theta)}{1} \cdot \dfrac{1}{1 + \cos(a\theta)} =$

$= \dfrac{1}{b}\lim\limits_{\theta \to 0} \dfrac{\sin(a\theta)}{\theta} \cdot \dfrac{a}{a} \cdot \dfrac{\sin(a\theta)}{1} \cdot \dfrac{1}{1 + \cos(a\theta)} = \dfrac{a}{b}\lim\limits_{\theta \to 0} \dfrac{\sin(a\theta)}{a\theta} \cdot \dfrac{\sin(a\theta)}{1} \cdot \dfrac{1}{1 + \cos(a\theta)} =$

$= \dfrac{a}{b} \cdot \lim\limits_{\theta \to 0} \dfrac{\sin(a\theta)}{a\theta} \cdot \lim\limits_{\theta \to 0} \sin(a\theta) \cdot \lim\limits_{\theta \to 0} \dfrac{1}{1 + \cos(a\theta)} =$

$= \dfrac{a}{b} \cdot \lim\limits_{\theta \to 0} \dfrac{\sin(a\theta)}{a\theta} \cdot \lim\limits_{\theta \to 0} \sin(a\theta) \cdot \dfrac{\lim\limits_{\theta \to 0} 1}{\lim\limits_{\theta \to 0}[1 + \cos(a\theta)]} = \dfrac{a}{b} \cdot 1 \cdot 0 \cdot \dfrac{1}{1 + 1} = 0.$

Tangent Group:

29. $\lim\limits_{\theta \to 0} \tan(\theta) = \lim\limits_{\theta \to 0} \dfrac{\sin(\theta)}{\cos(\theta)} = \dfrac{\lim\limits_{\theta \to 0} \sin(\theta)}{\lim\limits_{\theta \to 0} \cos(\theta)} = \dfrac{\sin(0)}{\cos(0)} = \dfrac{0}{1} = 0.$

30. $\lim\limits_{\theta \to 0} 3\tan(\theta) = \lim\limits_{\theta \to 0} \dfrac{3\sin(\theta)}{\cos(\theta)} = \dfrac{\lim\limits_{\theta \to 0} 3\sin(\theta)}{\lim\limits_{\theta \to 0} \cos(\theta)} = \dfrac{3\lim\limits_{\theta \to 0} \sin(\theta)}{\lim\limits_{\theta \to 0} \cos(\theta)} = \dfrac{3\sin(0)}{\cos(0)} = \dfrac{3 \cdot 0}{1} = 0.$

31. $\lim\limits_{\theta \to 0} \dfrac{\tan(\theta)}{\theta} = \lim\limits_{\theta \to 0} \dfrac{\left(\dfrac{\sin(\theta)}{\cos(\theta)}\right)}{\left(\dfrac{\theta}{1}\right)} = \lim\limits_{\theta \to 0} \dfrac{\sin(\theta)}{\cos(\theta)} \cdot \dfrac{1}{\theta} = \lim\limits_{\theta \to 0} \dfrac{\sin(\theta)}{\theta} \cdot \dfrac{1}{\cos(\theta)} =$

$= \lim\limits_{\theta \to 0} \dfrac{\sin(\theta)}{\theta} \cdot \lim\limits_{\theta \to 0} \dfrac{1}{\cos(\theta)} = 1 \cdot \dfrac{1}{\cos(0)} = \dfrac{1}{1} = 1.$

32. $\lim\limits_{\theta \to 0} \dfrac{\theta}{\tan(\theta)} = \lim\limits_{\theta \to 0} \dfrac{1}{\left(\dfrac{\tan(\theta)}{\theta}\right)} = \dfrac{\lim\limits_{\theta \to 0} 1}{\lim\limits_{\theta \to 0} \left(\dfrac{\tan(\theta)}{\theta}\right)} \overset{\#31}{=} \dfrac{1}{1} = 1.$

15

33. $\displaystyle\lim_{\theta \to 0} \frac{\tan(3\theta)}{\theta} = \lim_{\theta \to 0} \frac{\tan(3\theta)}{\theta} \cdot \frac{3}{3} = \lim_{\theta \to 0} \frac{\tan(3\theta)}{3\theta} \cdot \frac{3}{1} = \lim_{\theta \to 0} \frac{\tan(3\theta)}{3\theta} \cdot \lim 3 = \lim_{\theta \to 0} \frac{\tan(3\theta)}{3\theta} \cdot 3 \overset{\#31}{=}$

$\overset{\#31}{=} 1 \cdot 3 = 3.$

34. $\displaystyle\lim_{\theta \to 0} \theta \cot(\theta) = \lim_{\theta \to 0} \frac{\theta}{\tan(\theta)} \overset{\#32}{=} 1.$

35. $\displaystyle\lim_{\theta \to 0} \frac{\tan(3\theta)}{\tan(4\theta)} = \lim_{\theta \to 0} \frac{\left(\frac{\sin(3\theta)}{\cos(3\theta)}\right)}{\left(\frac{\sin(4\theta)}{\cos(4\theta)}\right)} = \lim_{\theta \to 0} \frac{\sin(3\theta)}{\cos(3\theta)} \cdot \frac{\cos(4\theta)}{\sin(4\theta)} = \lim_{\theta \to 0} \frac{\sin(3\theta)}{1} \cdot \frac{1}{\sin(4\theta)} \cdot \frac{\cos(4\theta)}{\cos(3\theta)} =$

$\displaystyle = \lim_{\theta \to 0} \frac{\sin(3\theta)}{1} \cdot \frac{3\theta}{3\theta} \cdot \frac{1}{\sin(4\theta)} \cdot \frac{4\theta}{4\theta} \cdot \frac{\cos(4\theta)}{\cos(3\theta)} = \lim_{\theta \to 0} \frac{\sin(3\theta)}{3\theta} \cdot \frac{4\theta}{\sin(4\theta)} \cdot \frac{3\theta}{4\theta} \cdot \frac{\cos(4\theta)}{\cos(3\theta)} =$

$\displaystyle = \lim_{\theta \to 0} \frac{\sin(3\theta)}{3\theta} \cdot \lim_{\theta \to 0} \frac{4\theta}{\sin(4\theta)} \cdot \lim_{\theta \to 0} \frac{3}{4} \cdot \lim_{\theta \to 0} \frac{\cos(4\theta)}{\cos(3\theta)} = 1 \cdot 1 \cdot \frac{3}{4} \cdot \frac{\displaystyle\lim_{\theta \to 0} \cos(4\theta)}{\displaystyle\lim_{\theta \to 0} \cos(3\theta)} = \frac{3}{4} \cdot \frac{\cos(0)}{\cos(0)}$

$\displaystyle = \frac{3}{4} \cdot \frac{1}{1} = \frac{3}{4}.$

Alternate Solution:

$\displaystyle\lim_{\theta \to 0} \frac{\tan(3\theta)}{\tan(4\theta)} = \lim_{\theta \to 0} \frac{\tan(3\theta)}{1} \cdot \frac{1}{\tan(4\theta)} = \lim_{\theta \to 0} \frac{\tan(3\theta)}{3\theta} \cdot \frac{4\theta}{\tan(4\theta)} \cdot \frac{3}{4} =$

$\displaystyle = \lim_{\theta \to 0} \frac{\tan(3\theta)}{3\theta} \cdot \lim_{\theta \to 0} \frac{4\theta}{\tan(4\theta)} \cdot \lim_{\theta \to 0} \frac{3}{4} \overset{\#31\ \&\ \#32}{=} 1 \cdot 1 \cdot \frac{3}{4} = \frac{3}{4}.$

Let a and b be nonzero numbers.

36. $\displaystyle\lim_{\theta\to 0}\frac{\tan(a\theta)}{\tan(b\theta)}=\lim_{\theta\to 0}\frac{\left(\dfrac{\sin(a\theta)}{\cos(b\theta)}\right)}{\left(\dfrac{\sin(a\theta)}{\cos(b\theta)}\right)}=\lim_{\theta\to 0}\frac{\sin(a\theta)}{\cos(a\theta)}\cdot\frac{\cos(b\theta)}{\sin(b\theta)}=\lim_{\theta\to 0}\frac{\sin(a\theta)}{1}\cdot\frac{1}{\sin(b\theta)}\cdot\frac{\cos(b\theta)}{\cos(a\theta)}=$

$\displaystyle=\lim_{\theta\to 0}\frac{\sin(a\theta)}{1}\cdot\frac{a\theta}{a\theta}\cdot\frac{1}{\sin(b\theta)}\cdot\frac{b\theta}{b\theta}\cdot\frac{\cos(b\theta)}{\cos(a\theta)}=\lim_{\theta\to 0}\frac{\sin(a\theta)}{a\theta}\cdot\frac{b\theta}{\sin(b\theta)}\cdot\frac{a\theta}{b\theta}\cdot\frac{\cos(b\theta)}{\cos(a\theta)}=$

$\displaystyle=\lim_{\theta\to 0}\frac{\sin(a\theta)}{a\theta}\cdot\lim_{\theta\to 0}\frac{b\theta}{\sin(b\theta)}\cdot\lim_{\theta\to 0}\frac{a}{b}\cdot\lim_{\theta\to 0}\frac{\cos(b\theta)}{\cos(a\theta)}=1\cdot 1\cdot\frac{a}{b}\cdot\frac{\lim\limits_{\theta\to 0}\cos(b\theta)}{\lim\limits_{\theta\to 0}\cos(a\theta)}=\frac{a}{b}\cdot\frac{\cos(0)}{\cos(0)}$

$\displaystyle=\frac{a}{b}\cdot\frac{1}{1}=\frac{a}{b}.$

Alternate Solution:

$\displaystyle\lim_{\theta\to 0}\frac{\tan(a\theta)}{\tan(b\theta)}=\lim_{\theta\to 0}\frac{\tan(a\theta)}{1}\cdot\frac{1}{\tan(b\theta)}=\lim_{\theta\to 0}\frac{\tan(a\theta)}{a\theta}\cdot\frac{b\theta}{\tan(a\theta)}\cdot\frac{a}{b}=$

$\displaystyle=\lim_{\theta\to 0}\frac{\tan(a\theta)}{a\theta}\cdot\lim_{\theta\to 0}\frac{b\theta}{\tan(b\theta)}\cdot\lim_{\theta\to 0}\frac{a}{b}\overset{\text{\#31 \& \#32}}{=}1\cdot 1\cdot\frac{a}{b}=\frac{a}{b}.$

Trig-Identity Heavy Group:

37. $\displaystyle\lim_{\theta\to 0}\frac{\theta}{\cos\left(\frac{\pi}{2}-\theta\right)}=\lim_{\theta\to 0}\frac{\theta}{\sin(\theta)}\overset{\text{\#1}}{=}1.$

Alternative Solution:

$\displaystyle\lim_{\theta\to 0}\frac{\theta}{\cos\left(\frac{\pi}{2}-\theta\right)}=\lim_{\theta\to 0}\frac{\theta}{\cos\left(\frac{\pi}{2}\right)\cos(\theta)+\sin\left(\frac{\pi}{2}\right)\sin(\theta)}=\lim_{\theta\to 0}\frac{\theta}{(0)\cos(\theta)+(1)\sin(\theta)}=$

$\displaystyle=\lim_{\theta\to 0}\frac{\theta}{\sin(\theta)}\overset{\text{\#1}}{=}1.$

38. $\displaystyle\lim_{\theta\to 0}\frac{\cos\left(\theta+\frac{\pi}{2}\right)}{\theta}=\lim_{\theta\to 0}\frac{\cos(\theta)\cos\left(\frac{\pi}{2}\right)-\sin(\theta)\sin\left(\frac{\pi}{2}\right)}{\theta}=\lim_{\theta\to 0}\frac{\cos(\theta)(0)-\sin(\theta)(1)}{\theta}=$

$\displaystyle=\lim_{\theta\to 0}\frac{-\sin(\theta)}{\theta}=-\lim_{\theta\to 0}\frac{\sin(\theta)}{\theta}=-1.$

39. $\displaystyle\lim_{\theta\to0}\frac{\cos\left(\frac{\pi}{2}-\theta\right)}{\theta}=\lim_{\theta\to0}\frac{\sin(\theta)}{\theta}=1.$

Alternate solution:

$$\lim_{\theta\to0}\frac{\cos\left(\frac{\pi}{2}-\theta\right)}{\theta}=\lim_{\theta\to0}\frac{\cos\left(\frac{\pi}{2}\right)\cos(\theta)+\sin\left(\frac{\pi}{2}\right)\sin(\theta)}{\theta}=\lim_{\theta\to0}\frac{0\cdot\cos(\theta)+1\cdot\sin(\theta)}{\theta}=$$

$$\lim_{\theta\to0}\frac{0\cdot\cos(\theta)}{\theta}+\lim_{\theta\to0}\frac{\sin(\theta)}{\theta}=\lim_{\theta\to0}0+\lim_{\theta\to0}\frac{\sin(\theta)}{\theta}=0+1=1.$$

40. $\displaystyle\lim_{\theta\to0}\frac{\theta}{\sin\left(\theta+\frac{\pi}{2}\right)}=\lim_{\theta\to0}\frac{\theta}{\cos(\theta)}\overset{\#16}{=}1.$

Alternative Solution:

$$\lim_{\theta\to0}\frac{\theta}{\sin\left(\theta+\frac{\pi}{2}\right)}=\lim_{\theta\to0}\frac{\theta}{\sin(\theta)\cos\left(\frac{\pi}{2}\right)+\cos(\theta)\sin\left(\frac{\pi}{2}\right)}=\lim_{\theta\to0}\frac{\theta}{\sin(\theta)(0)+\cos(\theta)(1)}=$$

$$=\lim_{\theta\to0}\frac{\theta}{0+\cos(\theta)}=\lim_{\theta\to0}\frac{\theta}{\cos(\theta)}\overset{\#16}{=}1.$$

41. $\displaystyle\lim_{\theta\to0}\frac{\sin\left(\frac{\pi}{2}-\theta\right)\tan(\pi+\theta)}{\theta}=\lim_{\theta\to0}\frac{\cos(\theta)\tan(\theta)}{\theta}=\lim_{\theta\to0}\cos(\theta)\cdot\lim_{\theta\to0}\frac{\tan(\theta)}{\theta}=$$

$$=\cos(0)\lim_{\theta\to0}\frac{\tan(\theta)}{\theta}=1\cdot\lim_{\theta\to0}\frac{\tan(\theta)}{\theta}=1\cdot1=1.$$

Alternate solution:

$$\lim_{\theta\to0}\frac{\sin\left(\frac{\pi}{2}-\theta\right)\tan(\pi+\theta)}{\theta}=\lim_{\theta\to0}\frac{[\sin\left(\frac{\pi}{2}\right)\cos(\theta)-\cos\left(\frac{\pi}{2}\right)\sin(\theta)][\frac{\sin(\pi+\theta)}{\cos(\pi+\theta)}]}{\theta}=$$

$$=\lim_{\theta\to0}\frac{[1\cdot\cos(\theta)-0\cdot\sin(\theta)]\left[\frac{\sin(\pi)\cos(\theta)+\cos(\pi)\sin(\theta)}{\cos(\pi)\cos(\theta)-\sin(\pi)\sin(\theta)}\right]}{\theta}=$$

$$=\lim_{\theta\to0}\frac{\cos(\theta)\cdot\frac{0\cdot\cos(\theta)+(-1)\cdot\sin(\theta)}{(-1)\cdot\cos(\theta)-0\cdot\sin(\theta)}}{\theta}=\lim_{\theta\to0}\frac{\cos(\theta)\cdot\frac{0-\sin(\theta)}{-\cos(\theta)-0}}{\theta}=$$

$$=\lim_{\theta\to0}\frac{\cos(\theta)\cdot\frac{-\sin(\theta)}{-\cos(\theta)}}{\theta}=\lim_{\theta\to0}\frac{\sin(\theta)}{\theta}=1.$$

42. $\lim_{\theta \to 0} \sin\left(\frac{\pi}{2} + \theta\right) \cos\left(\frac{\pi}{2} + \theta\right) = \lim_{\theta \to 0} \cos(\theta)[-\sin(\theta)] = \cos(0)[-\sin(0)] = 1 \cdot 0 = 0.$

Alternate solution:

$\lim_{\theta \to 0} \sin\left(\frac{\pi}{2} + \theta\right) \cos\left(\frac{\pi}{2} + \theta\right) =$

$= \lim_{\theta \to 0} \left[\{\sin\left(\frac{\pi}{2}\right) \cos(\theta) + \cos\left(\frac{\pi}{2}\right) \sin(\theta)\}\{\cos\left(\frac{\pi}{2}\right) \cos(\theta) - \sin\left(\frac{\pi}{2}\right) \sin(\theta)\}\right] =$

$= \lim_{\theta \to 0} [\{\cos(\theta) + 0\}\{0 - \sin(\theta)\}] = \lim_{\theta \to 0} \cos(\theta)[-\sin(\theta)] = \cos(0)[-\sin(0)] = 1 \cdot 0 = 0.$

Let a and b be nonzero numbers.

43. $\lim_{\theta \to 0} \dfrac{\cos\left(\frac{\pi}{2} - a\theta\right)}{b\theta} = \lim_{\theta \to 0} \dfrac{\sin(a\theta)}{b\theta} = \dfrac{1}{b} \lim_{\theta \to 0} \dfrac{\sin(a\theta)}{\theta} = \dfrac{1}{b} \lim_{\theta \to 0} \dfrac{\sin(a\theta)}{\theta} \cdot \dfrac{a}{a} = \dfrac{a}{b} \lim_{\theta \to 0} \dfrac{\sin(a\theta)}{a\theta} =$

$= \dfrac{a}{b} \cdot 1 = \dfrac{a}{b}.$

Alternate Solution:

$\lim_{\theta \to 0} \dfrac{\cos\left(\frac{\pi}{2} - a\theta\right)}{b\theta} = \lim_{\theta \to 0} \dfrac{\cos\left(\frac{\pi}{2}\right)\cos(a\theta) + \sin\left(\frac{\pi}{2}\right)\sin(a\theta)}{b\theta} =$

$= \lim_{\theta \to 0} \dfrac{(0)\cos(a\theta) + (1)\sin(a\theta)}{b\theta} = \lim_{\theta \to 0} \dfrac{0 + (1)\sin(a\theta)}{b\theta} = \dfrac{1}{b} \lim_{\theta \to 0} \dfrac{\sin(a\theta)}{\theta} =$

$= \dfrac{1}{b} \lim_{\theta \to 0} \dfrac{\sin(a\theta)}{\theta} \cdot \dfrac{a}{a} = \dfrac{a}{b} \lim_{\theta \to 0} \dfrac{\sin(a\theta)}{a\theta} = \dfrac{a}{b} \cdot 1 = \dfrac{a}{b}.$

Let a and b be nonzero numbers.

44. $\lim_{\theta \to 0} \dfrac{\cos\left(\frac{\pi}{2} - a\theta\right)}{\cos\left(\frac{\pi}{2} - b\theta\right)} = \lim_{\theta \to 0} \dfrac{\sin(a\theta)}{\sin(b\theta)} = \lim_{\theta \to 0} \dfrac{\sin(a\theta)}{\sin(b\theta)} \cdot \dfrac{ab\theta}{ab\theta} = \dfrac{a}{b} \lim_{\theta \to 0} \dfrac{\sin(a\theta)}{a\theta} \cdot \dfrac{b\theta}{\sin(b\theta)} =$

$= \dfrac{a}{b} \cdot \lim_{\theta \to 0} \dfrac{\sin(a\theta)}{a\theta} \cdot \lim_{\theta \to 0} \dfrac{b\theta}{\sin(b\theta)} \overset{\#1}{=} \dfrac{a}{b} \cdot 1 \cdot 1 = \dfrac{a}{b}.$

Alternate Solution:

$\lim_{\theta \to 0} \dfrac{\cos\left(\frac{\pi}{2} - a\theta\right)}{\cos\left(\frac{\pi}{2} - b\theta\right)} = \lim_{\theta \to 0} \dfrac{\cos\left(\frac{\pi}{2}\right)\cos(a\theta) + \sin\left(\frac{\pi}{2}\right)\sin(a\theta)}{\cos\left(\frac{\pi}{2}\right)\cos(b\theta) + \sin\left(\frac{\pi}{2}\right)\sin(b\theta)} =$

$= \lim_{\theta \to 0} \dfrac{(0)\cos(a\theta) + (1)\sin(a\theta)}{(0)\cos(b\theta) + (1)\sin(b\theta)} = \lim_{\theta \to 0} \dfrac{0 + \sin(a\theta)}{0 + \sin(b\theta)} = \lim_{\theta \to 0} \dfrac{\sin(a\theta)}{\sin(b\theta)} = \lim_{\theta \to 0} \dfrac{\sin(a\theta)}{\sin(b\theta)} \cdot \dfrac{ab\theta}{ab\theta} =$

$= \dfrac{a}{b} \lim_{\theta \to 0} \dfrac{\sin(a\theta)}{a\theta} \cdot \dfrac{b\theta}{\sin(b\theta)} = \dfrac{a}{b} \cdot \lim_{\theta \to 0} \dfrac{\sin(a\theta)}{a\theta} \cdot \lim_{\theta \to 0} \dfrac{b\theta}{\sin(b\theta)} \overset{\#1}{=} \dfrac{a}{b} \cdot 1 \cdot 1 = \dfrac{a}{b}.$

Let a and b be nonzero numbers.

45. $\lim\limits_{\theta \to 0} \dfrac{\sin\left(\frac{\pi}{2} - a\theta\right)}{\sin\left(\frac{\pi}{2} - b\theta\right)} = \lim\limits_{\theta \to 0} \dfrac{\cos(a\theta)}{\cos(b\theta)} = \dfrac{\lim\limits_{\theta \to 0} \cos(a\theta)}{\lim\limits_{\theta \to 0} \cos(b\theta)} = \dfrac{\cos(0)}{\cos(0)} = \dfrac{1}{1} = 1.$

Alternate Solution:

$\lim\limits_{\theta \to 0} \dfrac{\sin\left(\frac{\pi}{2} - a\theta\right)}{\sin\left(\frac{\pi}{2} - b\theta\right)} = \lim\limits_{\theta \to 0} \dfrac{\sin\left(\frac{\pi}{2}\right)\cos(a\theta) - \cos\left(\frac{\pi}{2}\right)\sin(a\theta)}{\sin\left(\frac{\pi}{2}\right)\cos(b\theta) - \cos\left(\frac{\pi}{2}\right)\sin(b\theta)} =$

$\lim\limits_{\theta \to 0} \dfrac{(1)\cos(a\theta) - (0)\sin(a\theta)}{(1)\cos(b\theta) - (0)\sin(b\theta)} = \lim\limits_{\theta \to 0} \dfrac{\cos(a\theta)}{\cos(b\theta)} = \dfrac{\lim\limits_{\theta \to 0} \cos(a\theta)}{\lim\limits_{\theta \to 0} \cos(b\theta)} = \dfrac{\cos(0)}{\cos(0)} = \dfrac{1}{1} = 1.$

46. $\lim\limits_{\theta \to 0}[\sin(\pi - \theta) + \tan(\pi - \theta)] = \lim\limits_{\theta \to 0}[\sin(\theta) + \tan(\theta)] = \sin(0) + \tan(0) = 0 + 0 = 0.$

Alternate solution:

$\lim\limits_{\theta \to 0}[\sin(\pi - \theta) + \tan(\pi - \theta)] = \lim\limits_{\theta \to 0}\left[\sin(\pi)\cos(\theta) - \cos(\pi)\sin(\theta) + \dfrac{\tan(\pi) + \tan(\theta)}{1 - \tan(\pi)\tan(\theta)}\right] =$

$\lim\limits_{\theta \to 0}\left[0 \cdot \cos(\theta) - (-1)\sin(\theta) + \dfrac{0 + \tan(\theta)}{1 - 0 \cdot \tan(\theta)}\right] = \lim\limits_{\theta \to 0}[\sin(\theta) + \tan(\theta)] =$

$= \sin(0) + \tan(0) = 0 + 0 = 0.$

Miscellaneous Group:

Let a and b be nonzero numbers.

47. $\lim\limits_{\theta \to 0} \dfrac{\tan(a\theta)}{\sin(b\theta)} = \lim\limits_{\theta \to 0} \dfrac{\sin(a\theta)}{\cos(a\theta)} \cdot \dfrac{1}{\sin(b\theta)} = \lim\limits_{\theta \to 0} \dfrac{\sin(a\theta)}{1} \cdot \dfrac{1}{\cos(a\theta)} \cdot \dfrac{1}{\sin(b\theta)} =$

$= \lim\limits_{\theta \to 0} \dfrac{\sin(a\theta)}{1} \cdot \dfrac{a\theta}{a\theta \cos(a\theta)} \cdot \dfrac{1}{\sin(b\theta)} \cdot \dfrac{b\theta}{b\theta} = \lim\limits_{\theta \to 0} \dfrac{\sin(a\theta)}{a\theta} \cdot \dfrac{1}{\cos(a\theta)} \cdot \dfrac{b\theta}{\sin(b\theta)} \cdot \dfrac{a\theta}{b\theta} =$

$= \lim\limits_{\theta \to 0} \dfrac{\sin(a\theta)}{a\theta} \cdot \dfrac{1}{\cos(a\theta)} \cdot \dfrac{b\theta}{\sin(b\theta)} \cdot \dfrac{a}{b} = \dfrac{a}{b} \cdot \lim\limits_{\theta \to 0} \dfrac{\sin(a\theta)}{a\theta} \cdot \lim\limits_{\theta \to 0} \dfrac{1}{\cos(a\theta)} \cdot \lim\limits_{\theta \to 0} \dfrac{b\theta}{\sin(b\theta)} \overset{\#1}{=}$

$\overset{\#1}{=} \dfrac{a}{b} \cdot 1 \cdot \dfrac{\lim\limits_{\theta \to 0} 1}{\lim\limits_{\theta \to 0} \cos(a\theta)} \cdot 1 = \dfrac{a}{b} \cdot 1 \cdot \dfrac{1}{\cos(0)} \cdot 1 = \dfrac{a}{b} \cdot \dfrac{1}{1} = \dfrac{a}{b}.$

48. $\displaystyle\lim_{\theta\to 0}\frac{\sin(\theta)}{1-\cos(\theta)}=\lim_{\theta\to 0}\frac{\sin(\theta)}{1}\cdot\frac{1}{1-\cos(\theta)}=\lim_{\theta\to 0}\frac{\sin(\theta)}{1}\cdot\frac{1}{1-\cos(\theta)}\cdot\frac{\theta}{\theta}=$

$\displaystyle=\lim_{\theta\to 0}\frac{\sin(\theta)}{\theta}\cdot\frac{\theta}{1-\cos(\theta)}=\lim_{\theta\to 0}\frac{\sin(\theta)}{\theta}\cdot\lim_{\theta\to 0}\frac{\theta}{1-\cos(\theta)}=1\cdot\lim_{\theta\to 0}\frac{\theta}{1-\cos(\theta)}=$

$\displaystyle=\lim_{\theta\to 0}\frac{\theta}{1-\cos(\theta)}\cdot\frac{1+\cos(\theta)}{1+\cos(\theta)}=\lim_{\theta\to 0}\frac{\theta[1+\cos(\theta)]}{1-\cos^2\theta}=\lim_{\theta\to 0}\frac{\theta[1+\cos(\theta)]}{\sin^2\theta}=$

$\displaystyle=\lim_{\theta\to 0}\frac{\theta}{\sin(\theta)}\cdot\frac{1+\cos(\theta)}{\sin(\theta)}=\lim_{\theta\to 0}\frac{\theta}{\sin(\theta)}\cdot\lim_{\theta\to 0}\frac{1+\cos(\theta)}{\sin(\theta)}\overset{\#1}{=}1\cdot\lim_{\theta\to 0}\frac{1+\cos(\theta)}{\sin(\theta)}=$

$\displaystyle=\lim_{\theta\to 0}\frac{1+\cos(\theta)}{\sin(\theta)}=$ Does not exist.

Remark: Note that the numerator is approaching the fixed number 2 while the denominator is approaching 0. Furthermore, $\displaystyle\lim_{\theta\to 0^+}\frac{1+\cos(\theta)}{\sin(\theta)}=+\infty$ while $\displaystyle\lim_{\theta\to 0^-}\frac{1+\cos(\theta)}{\sin(\theta)}=-\infty$.

49. $\displaystyle\lim_{\theta\to 0}\frac{5\theta+\sin(3\theta)}{\tan(4\theta)-7\theta\cos(6\theta)}=\lim_{\theta\to 0}\frac{5\theta+\sin(3\theta)}{\dfrac{\sin(4\theta)}{\cos(4\theta)}-7\theta\cos(6\theta)}=$

$\displaystyle=\lim_{\theta\to 0}\frac{5\theta+\sin(3\theta)}{\dfrac{\sin(4\theta)}{1}\cdot\dfrac{1}{\cos(4\theta)}-7\theta\cos(6\theta)}=\lim_{\theta\to 0}\frac{5\theta+\sin(3\theta)}{\dfrac{\sin(4\theta)}{1}\cdot\dfrac{1}{\cos(4\theta)}-7\theta\cos(6\theta)}\cdot\frac{\left(\dfrac{1}{\theta}\right)}{\left(\dfrac{1}{\theta}\right)}=$

$\displaystyle=\lim_{\theta\to 0}\frac{\dfrac{5\theta}{\theta}+\dfrac{\sin(3\theta)}{\theta}}{\dfrac{\sin(4\theta)}{\theta}\cdot\dfrac{1}{\cos(4\theta)}-\dfrac{7\theta\cos(6\theta)}{\theta}}=\lim_{\theta\to 0}\frac{5+\dfrac{\sin(3\theta)}{\theta}\cdot\dfrac{3}{3}}{\dfrac{\sin(4\theta)}{\theta}\cdot\dfrac{4}{4}\cdot\dfrac{1}{\cos(4\theta)}-\dfrac{7\cos(6\theta)}{1}}=$

$\displaystyle=\lim_{\theta\to 0}\frac{5+\dfrac{\sin(3\theta)}{3\theta}\cdot\dfrac{3}{1}}{\dfrac{\sin(4\theta)}{4\theta}\cdot\dfrac{4}{1}\cdot\dfrac{1}{\cos(4\theta)}-\dfrac{7\cos(6\theta)}{1}}=\frac{\displaystyle\lim_{\theta\to 0}\left(5+\dfrac{\sin(3\theta)}{3\theta}\cdot\dfrac{3}{1}\right)}{\displaystyle\lim_{\theta\to 0}\left(\dfrac{\sin(4\theta)}{4\theta}\cdot\dfrac{4}{1}\cdot\dfrac{1}{\cos(4\theta)}-\dfrac{7\cos(6\theta)}{1}\right)}=$

$\displaystyle=\frac{\displaystyle\lim_{\theta\to 0}5+\lim_{\theta\to 0}\frac{\sin(3\theta)}{3\theta}\cdot\lim_{\theta\to 0}\frac{3}{1}}{\displaystyle\lim_{\theta\to 0}\frac{\sin(4\theta)}{4\theta}\cdot\lim_{\theta\to 0}\frac{4}{1}\cdot\lim_{\theta\to 0}\frac{1}{\cos(4\theta)}-\lim_{\theta\to 0}\frac{7\cos(6\theta)}{1}}=$

$\displaystyle=\frac{\displaystyle\lim_{\theta\to 0}5+\lim_{\theta\to 0}\frac{\sin(3\theta)}{3\theta}\cdot\lim_{\theta\to 0}\frac{3}{1}}{\displaystyle\lim_{\theta\to 0}\frac{\sin(4\theta)}{4\theta}\cdot\lim_{\theta\to 0}\frac{4}{1}\cdot\frac{\displaystyle\lim_{\theta\to 0}1}{\displaystyle\lim_{\theta\to 0}\cos(4\theta)}-\frac{\displaystyle\lim_{\theta\to 0}7\cos(6\theta)}{\displaystyle\lim_{\theta\to 0}1}}=\frac{5+1\cdot 3}{1\cdot 4\cdot\dfrac{1}{\cos(0)}-\dfrac{7\cos(0)}{1}}=$

$\displaystyle=\frac{5+1\cdot 3}{1\cdot 4\cdot 1-7}=\frac{5+3}{4-7}=\frac{8}{-3}.$

50.

$$\lim_{\theta \to 0} \frac{1 - \cos^3(\theta)}{\sin^2(\theta)} \overset{\substack{\text{"difference of} \\ \text{cubes" factoring} \\ \text{formula}}}{=} \lim_{\theta \to 0} \frac{(1 - \cos(\theta))(1 + \cos(\theta) + \cos^2(\theta))}{1 - \cos^2(\theta)} =$$

$$= \lim_{\theta \to 0} \frac{(1 - \cos(\theta))(1 + \cos(\theta) + \cos^2(\theta))}{(1 + \cos(\theta))(1 - \cos(\theta))} = \lim_{\theta \to 0} \frac{1 + \cos(\theta) + \cos^2(\theta)}{1 + \cos(\theta)} =$$

$$= \frac{\lim_{\theta \to 0}(1 + \cos(\theta) + \cos^2(\theta))}{\lim_{\theta \to 0}(1 + \cos(\theta))} = \frac{\lim_{\theta \to 0} 1 + \lim_{\theta \to 0} \cos(\theta) + \lim_{\theta \to 0} \cos(\theta) \cdot \lim_{\theta \to 0} \cos(\theta)}{\lim_{\theta \to 0} 1 + \lim_{\theta \to 0} \cos(\theta)} =$$

$$= \frac{1 + \cos(0) + \cos(0) \cdot \cos(0)}{1 + \cos(0)} = \frac{1 + 1 + 1 \cdot 1}{1 + 1} = \frac{3}{2}.$$

51.

$$\lim_{\theta \to 0} \frac{2 + \sin(\theta)}{3 + \theta} = \frac{\lim_{\theta \to 0}[2 + \sin(\theta)]}{\lim_{\theta \to 0}[3 + \theta]} = \frac{2 + \sin(0)}{3 + 0} = \frac{2 + 0}{3 + 0} = \frac{2}{3}.$$

52.

$$\lim_{\theta \to 0} \frac{\theta^2 + 1}{\theta + \cos(\theta)} = \frac{\lim_{\theta \to 0}[\theta^2 + 1]}{\lim_{\theta \to 0}[\theta + \cos(\theta)]} = \frac{0^2 + 1}{0 + \cos(0)} = \frac{1}{0 + 1} = 1.$$

53.

$$\lim_{\theta \to 0} \frac{\theta + \tan(\theta)}{\sin(\theta)} = \lim_{\theta \to 0} \frac{\theta + \frac{\sin(\theta)}{\cos(\theta)}}{\sin(\theta)} = \lim_{\theta \to 0} \frac{\theta + \frac{\sin(\theta)}{\cos(\theta)}}{\frac{\sin(\theta)}{1}} = \lim_{\theta \to 0} \frac{1}{\sin(\theta)}\left(\theta + \frac{\sin(\theta)}{\cos(\theta)}\right) =$$

$$= \lim_{\theta \to 0} \left(\frac{\theta}{\sin(\theta)} + \frac{1}{\cos(\theta)}\right) = \lim_{\theta \to 0} \frac{\theta}{\sin(\theta)} + \lim_{\theta \to 0} \frac{1}{\cos(\theta)} = \lim_{\theta \to 0} \frac{\theta}{\sin(\theta)} + \frac{\lim_{\theta \to 0} 1}{\lim_{\theta \to 0} \cos(\theta)} \overset{\#1}{=}$$

$$\overset{\#1}{=} 1 + \frac{1}{\cos(0)} = 1 + \frac{1}{1} = 1 + 1 = 2.$$

54.

$$\lim_{\theta \to 0} \frac{1 - \cos(3\theta)}{\theta \sin(\theta)} = \lim_{\theta \to 0} \frac{1 - \cos(3\theta)}{\theta \sin(\theta)} \cdot \frac{1 + \cos(3\theta)}{1 + \cos(3\theta)} = \lim_{\theta \to 0} \frac{1 - \cos^2(3\theta)}{\theta \sin(\theta)\,(1 + \cos(3\theta))} =$$

$$= \lim_{\theta \to 0} \frac{\sin^2(3\theta)}{\theta \sin(\theta)\,(1 + \cos(3\theta))} = \lim_{\theta \to 0} \frac{\sin(3\theta)}{\theta} \cdot \frac{\sin(3\theta)}{1} \cdot \frac{1}{\sin(\theta)} \cdot \frac{1}{1 + \cos(3\theta)} =$$

$$= \lim_{\theta \to 0} \frac{\sin(3\theta)}{\theta} \cdot \frac{3}{3} \cdot \frac{\sin(3\theta)}{1} \cdot \frac{3\theta}{3\theta} \cdot \frac{1}{\sin(\theta)} \cdot \frac{1}{1 + \cos(3\theta)} =$$

$$= \lim_{\theta \to 0} \frac{\sin(3\theta)}{3\theta} \cdot \frac{\sin(3\theta)}{3\theta} \cdot \frac{\theta}{\sin(\theta)} \cdot \frac{9}{1 + \cos(3\theta)} =$$

$$= \lim_{\theta \to 0} \frac{\sin(3\theta)}{3\theta} \cdot \lim_{\theta \to 0} \frac{\sin(3\theta)}{3\theta} \cdot \lim_{\theta \to 0} \frac{\theta}{\sin(\theta)} \cdot \lim_{\theta \to 0} \frac{9}{1 + \cos(3\theta)} =$$

$$= \lim_{\theta \to 0} \frac{\sin(3\theta)}{3\theta} \cdot \lim_{\theta \to 0} \frac{\sin(3\theta)}{3\theta} \cdot \lim_{\theta \to 0} \frac{\theta}{\sin(\theta)} \cdot \frac{\lim\limits_{\theta \to 0} 9}{\lim\limits_{\theta \to 0}(1 + \cos(3\theta))} \overset{\#1}{=} 1 \cdot 1 \cdot 1 \cdot \frac{9}{1 + \cos(0)} =$$

$$= \frac{9}{1 + 1} = \frac{9}{2}.$$

55. $$\lim_{\theta \to 0} \frac{1 - \cos(2\theta)}{\theta \sin(\theta)} = \lim_{\theta \to 0} \frac{1 - \cos(2\theta)}{\theta \sin(\theta)} \cdot \frac{1 + \cos(2\theta)}{1 + \cos(2\theta)} = \lim_{\theta \to 0} \frac{1 - \cos^2(2\theta)}{\theta \sin(\theta)(1 + \cos(2\theta))} =$$

$$= \lim_{\theta \to 0} \frac{\sin^2(2\theta)}{\theta \sin(\theta)(1 + \cos(2\theta))} = \lim_{\theta \to 0} \frac{\sin(2\theta)}{\theta} \cdot \frac{\sin(2\theta)}{1} \cdot \frac{1}{\sin(\theta)(1 + \cos(2\theta))} =$$

$$= \lim_{\theta \to 0} \frac{\sin(2\theta)}{\theta} \cdot \frac{2}{2} \cdot \frac{\sin(2\theta)}{1} \cdot \frac{2\theta}{2\theta} \cdot \frac{1}{\sin(\theta)(1 + \cos(2\theta))} =$$

$$= \lim_{\theta \to 0} \frac{\sin(2\theta)}{2\theta} \cdot \frac{\sin(2\theta)}{2\theta} \cdot \frac{\theta}{\sin(\theta)} \cdot \frac{4}{1 + \cos(2\theta)} =$$

$$= \lim_{\theta \to 0} \frac{\sin(2\theta)}{2\theta} \cdot \lim_{\theta \to 0} \frac{\sin(2\theta)}{2\theta} \cdot \lim_{\theta \to 0} \frac{\theta}{\sin(\theta)} \cdot \lim_{\theta \to 0} \frac{4}{1 + \cos(2\theta)} =$$

$$= \lim_{\theta \to 0} \frac{\sin(2\theta)}{2\theta} \cdot \lim_{\theta \to 0} \frac{\sin(2\theta)}{2\theta} \cdot \lim_{\theta \to 0} \frac{\theta}{\sin(\theta)} \cdot \frac{\lim\limits_{\theta \to 0} 4}{\lim\limits_{\theta \to 0}[1 + \cos(2\theta)]} \overset{\#1}{=} 1 \cdot 1 \cdot 1 \cdot \frac{4}{1 + \cos(0)} =$$

$$= \frac{4}{1 + 1} = \frac{4}{2} = 2.$$

Alternate solution:

$$\lim_{\theta \to 0} \frac{1 - \cos(2\theta)}{\theta \sin(\theta)} = \lim_{\theta \to 0} \frac{1 - \cos(2\theta)}{\theta \sin(\theta)} \cdot \frac{1 + \cos(2\theta)}{1 + \cos(2\theta)} = \lim_{\theta \to 0} \frac{1 - \cos^2(2\theta)}{\theta \sin(\theta)(1 + \cos(2\theta))} =$$

$$= \lim_{\theta \to 0} \frac{\sin^2(2\theta)}{\theta \sin(\theta)(1 + \cos(2\theta))} = \lim_{\theta \to 0} \frac{\sin(2\theta)}{\theta} \cdot \frac{\sin(2\theta)}{1} \cdot \frac{1}{\sin(\theta)(1 + \cos(2\theta))} =$$

$$= \lim_{\theta \to 0} \frac{2\sin(\theta)\cos(\theta)}{\theta} \cdot \frac{2\sin(\theta)\cos(\theta)}{1} \cdot \frac{1}{\sin(\theta)(1 + \cos(2\theta))} =$$

$$= 4 \lim_{\theta \to 0} \frac{\sin(\theta)\cos(\theta)}{\theta} \cdot \frac{\cos(\theta)}{1} \cdot \frac{1}{(1 + \cos(2\theta))} = 4 \lim_{\theta \to 0} \frac{\sin(\theta)}{\theta} \cdot \frac{\cos^2(\theta)}{1 + \cos(2\theta)} =$$

$$= 4 \lim_{\theta \to 0} \frac{\sin(\theta)}{\theta} \cdot \lim_{\theta \to 0} \frac{\cos^2(\theta)}{1 + \cos(2\theta)} = 4 \lim_{\theta \to 0} \frac{\sin(\theta)}{\theta} \cdot \frac{\lim\limits_{\theta \to 0} \cos^2(\theta)}{\lim\limits_{\theta \to 0}[1 + \cos(2\theta)]} = 4 \cdot 1 \cdot \frac{\cos^2(0)}{1 + \cos(0)} =$$

$$= \frac{4(1)}{1 + 1} = \frac{4}{2} = 2.$$

56. $\lim\limits_{\theta \to 0} \dfrac{1 - \cos(\theta)}{\theta \sin(\theta)} = \lim\limits_{\theta \to 0} \dfrac{1 - \cos(\theta)}{\theta \sin(\theta)} \cdot \dfrac{1 + \cos(\theta)}{1 + \cos(\theta)} = \lim\limits_{\theta \to 0} \dfrac{1 - \cos^2(\theta)}{\theta \sin(\theta)\,(1 + \cos(\theta))} =$

$= \lim\limits_{\theta \to 0} \dfrac{\sin^2(\theta)}{\theta \sin(\theta)\,(1 + \cos(\theta))} = \lim\limits_{\theta \to 0} \dfrac{\sin(\theta)}{\theta(1 + \cos(\theta))} = \lim\limits_{\theta \to 0} \dfrac{\sin(\theta)}{\theta} \cdot \dfrac{1}{1 + \cos(\theta)} =$

$= \lim\limits_{\theta \to 0} \dfrac{\sin(\theta)}{\theta} \cdot \lim\limits_{\theta \to 0} \dfrac{1}{1 + \cos(\theta)} = \lim\limits_{\theta \to 0} \dfrac{\sin(\theta)}{\theta} \cdot \dfrac{\lim\limits_{\theta \to 0} 1}{\lim\limits_{\theta \to 0}[1 + \cos(\theta)]} = 1 \cdot \dfrac{1}{1 + \cos(0)} = \dfrac{1}{1 + 1} = \dfrac{1}{2}.$

57. $\lim\limits_{\theta \to 0} \dfrac{\cos(\theta)}{\theta \csc(\theta)} = \lim\limits_{\theta \to 0} \dfrac{\cos(\theta)}{1} \cdot \dfrac{\sin(\theta)}{\theta} = \lim\limits_{\theta \to 0} \cos(\theta) \cdot \lim\limits_{\theta \to 0} \dfrac{\sin(\theta)}{\theta} = \cos(0) \cdot 1 = 1 \cdot 1 = 1.$

58. $\lim\limits_{\theta \to 0} \sin(2\theta) \cot(\theta) = \lim\limits_{\theta \to 0} \sin(2\theta) \cdot \dfrac{\cos(\theta)}{\sin(\theta)} = \lim\limits_{\theta \to 0} 2 \sin(\theta) \cos(\theta) \cdot \dfrac{\cos(\theta)}{\sin(\theta)} =$

$= \lim\limits_{\theta \to 0} 2 \cos(\theta) \cos(\theta) = 2 \cos(0) \cos(0) = 2 \cdot 1 \cdot 1 = 2.$

Alternate solution:

$\lim\limits_{\theta \to 0} \sin(2\theta) \cot(\theta) = \lim\limits_{\theta \to 0} \sin(2\theta) \cdot \dfrac{\cos(\theta)}{\sin(\theta)} = \lim\limits_{\theta \to 0} \dfrac{\sin(2\theta)}{1} \cdot \dfrac{\cos(\theta)}{1} \cdot \dfrac{1}{\sin(\theta)} =$

$= \lim\limits_{\theta \to 0} \dfrac{\sin(2\theta)}{1} \cdot \dfrac{2\theta}{2\theta} \cdot \dfrac{\cos(\theta)}{1} \cdot \dfrac{1}{\sin(\theta)} = \lim\limits_{\theta \to 0} \dfrac{2}{1} \cdot \dfrac{\sin(2\theta)}{2\theta} \cdot \dfrac{\cos(\theta)}{1} \cdot \dfrac{\theta}{\sin(\theta)} =$

$= \lim\limits_{\theta \to 0} \dfrac{2}{1} \cdot \lim\limits_{\theta \to 0} \dfrac{\sin(2\theta)}{2\theta} \cdot \lim\limits_{\theta \to 0} \dfrac{\cos(\theta)}{1} \cdot \lim\limits_{\theta \to 0} \dfrac{\theta}{\sin(\theta)} \overset{\#1}{=} 2 \cdot 1 \cdot \cos(0) \cdot 1 = 2 \cdot 1 \cdot 1 \cdot 1 = 2.$

59. $\lim\limits_{\theta \to 0} \dfrac{1 - \cos(\theta)}{\sin(\theta)} = \lim\limits_{\theta \to 0} \dfrac{1 - \cos(\theta)}{\sin(\theta)} \cdot \dfrac{1 + \cos(\theta)}{1 + \cos(\theta)} = \lim\limits_{\theta \to 0} \dfrac{1 - \cos^2(\theta)}{\sin(\theta)(1 + \cos(\theta))} =$

$= \lim\limits_{\theta \to 0} \dfrac{\sin^2(\theta)}{\sin(\theta)(1 + \cos(\theta))} = \lim\limits_{\theta \to 0} \dfrac{\sin(\theta)}{1 + \cos(\theta)} = \dfrac{\lim\limits_{\theta \to 0} \sin(\theta)}{\lim\limits_{\theta \to 0}(1 + \cos(\theta))} = \dfrac{\lim\limits_{\theta \to 0} \sin(\theta)}{\lim\limits_{\theta \to 0} 1 + \lim\limits_{\theta \to 0} \cos(\theta)} =$

$= \dfrac{0}{1 + \cos(0)} = \dfrac{0}{1 + 1} = \dfrac{0}{2} = 0.$

60. $\displaystyle \lim_{\theta \to 0} \frac{1 - \cos(\theta)}{\tan(\theta)} = \lim_{\theta \to 0} \frac{1 - \cos(\theta)}{\tan(\theta)} \cdot \frac{1 + \cos(\theta)}{1 + \cos(\theta)} = \lim_{\theta \to 0} \frac{1 - \cos^2(\theta)}{\tan(\theta)(1 + \cos(\theta))} =$

$\displaystyle = \lim_{\theta \to 0} \frac{\sin^2(\theta)}{\tan(\theta)(1 + \cos(\theta))} = \lim_{\theta \to 0} \frac{\sin^2(\theta)}{1} \cdot \frac{1}{\tan(\theta)} \cdot \frac{1}{1 + \cos(\theta)} =$

$\displaystyle = \lim_{\theta \to 0} \frac{\sin^2(\theta)}{1} \cdot \frac{\cot(\theta)}{1} \cdot \frac{1}{1 + \cos(\theta)} = \lim_{\theta \to 0} \frac{\sin^2(\theta)}{1} \cdot \frac{\cos(\theta)}{\sin(\theta)} \cdot \frac{1}{1 + \cos(\theta)} =$

$\displaystyle = \lim_{\theta \to 0} \frac{\sin(\theta)}{1} \cdot \frac{\cos(\theta)}{1} \cdot \frac{1}{1 + \cos(\theta)} = \lim_{\theta \to 0} \sin(\theta) \cdot \lim_{\theta \to 0} \cos(\theta) \cdot \lim_{\theta \to 0} \frac{1}{1 + \cos(\theta)} =$

$\displaystyle = \lim_{\theta \to 0} \sin(\theta) \cdot \lim_{\theta \to 0} \cos(\theta) \cdot \frac{\lim_{\theta \to 0} 1}{\lim_{\theta \to 0} [1 + \cos(\theta)]} = \sin(0) \cdot \cos(0) \cdot \frac{1}{1 + \cos(0)} =$

$\displaystyle = 0 \cdot 1 \cdot \frac{1}{1 + 1} = 0 \cdot 1 \cdot \frac{1}{2} = 0.$

61. $\displaystyle \lim_{\theta \to 0} \frac{\sin(\theta)}{\tan(\theta)} = \lim_{\theta \to 0} \frac{\sin(\theta)}{\left(\dfrac{\sin(\theta)}{\cos(\theta)}\right)} = \lim_{\theta \to 0} \frac{\left(\dfrac{\sin(\theta)}{1}\right)}{\left(\dfrac{\sin(\theta)}{\cos(\theta)}\right)} = \lim_{\theta \to 0} \frac{\sin(\theta)}{1} \cdot \frac{\cos(\theta)}{\sin(\theta)} = \lim_{\theta \to 0} \cos(\theta) =$

$\displaystyle = \cos(0) = 1.$

62. $\displaystyle \lim_{\theta \to 0} \frac{2\sin(\theta) - \sin(2\theta)}{\theta \cos(\theta)} = \lim_{\theta \to 0} \frac{2\sin(\theta) - 2\sin(\theta)\cos(\theta)}{\theta \cos(\theta)} = \lim_{\theta \to 0} \frac{2\sin(\theta)(1 - \cos(\theta))}{\theta \cos(\theta)} =$

$\displaystyle = \lim_{\theta \to 0} \frac{2}{1} \cdot \frac{\sin(\theta)}{\theta} \cdot \frac{1 - \cos(\theta)}{\cos(\theta)} = \lim_{\theta \to 0} \frac{2}{1} \cdot \lim_{\theta \to 0} \frac{\sin(\theta)}{\theta} \cdot \lim_{\theta \to 0} \frac{1 - \cos(\theta)}{\cos(\theta)} =$

$\displaystyle = 2 \cdot 1 \cdot \frac{\lim_{\theta \to 0} [1 - \cos(\theta)]}{\lim_{\theta \to 0} \cos(\theta)} = 2 \cdot \frac{1 - \cos(0)}{\cos(0)} = 2 \cdot \frac{1 - 1}{1} = 2 \cdot 0 = 0.$

63. $\displaystyle \lim_{\theta \to 0} \frac{\tan(\theta) - \sin(\theta)}{\theta \cos(\theta)} = \lim_{\theta \to 0} \frac{\dfrac{\sin(\theta)}{\cos(\theta)} - \dfrac{\sin(\theta)}{1}}{\theta \cos(\theta)} = \lim_{\theta \to 0} \frac{\left(\dfrac{\sin(\theta) - \cos(\theta)\sin(\theta)}{\cos(\theta)}\right)}{\left(\dfrac{\theta \cos(\theta)}{1}\right)} =$

$\displaystyle = \lim_{\theta \to 0} \frac{\sin(\theta) - \cos(\theta)\sin(\theta)}{\cos(\theta)} \cdot \frac{1}{\theta \cos(\theta)} = \lim_{\theta \to 0} \frac{\sin(\theta)(1 - \cos(\theta))}{\cos(\theta)} \cdot \frac{1}{\theta \cos(\theta)} =$

$\displaystyle = \lim_{\theta \to 0} \frac{\sin(\theta)}{\theta} \cdot \frac{1 - \cos(\theta)}{\cos^2(\theta)} = \lim_{\theta \to 0} \frac{\sin(\theta)}{\theta} \cdot \lim_{\theta \to 0} \frac{1 - \cos(\theta)}{\cos^2(\theta)} = \lim_{\theta \to 0} \frac{\sin(\theta)}{\theta} \cdot \frac{\lim_{\theta \to 0}(1 - \cos(\theta))}{\lim_{\theta \to 0} \cos^2(\theta)} =$

$\displaystyle = 1 \cdot \frac{1 - \cos(0)}{\cos^2(0)} = 1 \cdot \frac{1 - 1}{1} = 1 \cdot 0 = 0.$

64.
$$\lim_{\theta \to 0} \frac{\csc(\theta) - \cot(\theta)}{\sin(\theta)} = \lim_{\theta \to 0} \frac{\dfrac{1}{\sin(\theta)} - \dfrac{\cos(\theta)}{\sin(\theta)}}{\dfrac{\sin(\theta)}{1}} = \lim_{\theta \to 0} \frac{\left(\dfrac{1-\cos(\theta)}{\sin(\theta)}\right)}{\left(\dfrac{\sin(\theta)}{1}\right)} = \lim_{\theta \to 0} \frac{1-\cos(\theta)}{\sin(\theta)} \cdot \frac{1}{\sin(\theta)} =$$

$$= \lim_{\theta \to 0} \frac{1-\cos(\theta)}{\sin^2(\theta)} = \lim_{\theta \to 0} \frac{1-\cos(\theta)}{\sin^2(\theta)} \cdot \frac{1+\cos(\theta)}{1+\cos(\theta)} = \lim_{\theta \to 0} \frac{1-\cos^2(\theta)}{\sin^2(\theta)\,(1+\cos(\theta))} =$$

$$= \lim_{\theta \to 0} \frac{\sin^2(\theta)}{\sin^2(\theta)\,(1+\cos(\theta))} = \lim_{\theta \to 0} \frac{1}{1+\cos(\theta)} = \frac{\displaystyle\lim_{\theta \to 0} 1}{\displaystyle\lim_{\theta \to 0}[1+\cos(\theta)]} = \frac{1}{1+\cos(0)} =$$

$$= \frac{1}{1+1} = \frac{1}{2}.$$

65.
$$\lim_{\theta \to 0} \frac{2\theta + 1 - \cos(\theta)}{3\theta} = \lim_{\theta \to 0} \left(\frac{2\theta}{3\theta} + \frac{1-\cos(\theta)}{3\theta}\right) = \lim_{\theta \to 0} \frac{2}{3} + \lim_{\theta \to 0} \frac{1-\cos(\theta)}{3\theta} =$$

$$= \frac{2}{3} + \frac{1}{3}\lim_{\theta \to 0} \frac{1-\cos(\theta)}{\theta} \overset{\#19}{=} \frac{2}{3} + \frac{1}{3}(0) = \frac{2}{3}.$$

66.
$$\lim_{\theta \to 0} \frac{\sin^3(\theta)}{(2\theta)^3} = \lim_{\theta \to 0} \frac{\sin^3(\theta)}{8\theta^3} = \frac{1}{8}\lim_{\theta \to 0} \frac{\sin^3(\theta)}{\theta^3} = \frac{1}{8}\lim_{\theta \to 0} \left(\frac{\sin(\theta)}{\theta} \cdot \frac{\sin(\theta)}{\theta} \cdot \frac{\sin(\theta)}{\theta}\right) =$$

$$= \frac{1}{8}\left(\lim_{\theta \to 0} \frac{\sin(\theta)}{\theta} \cdot \lim_{\theta \to 0} \frac{\sin(\theta)}{\theta} \cdot \lim_{\theta \to 0} \frac{\sin(\theta)}{\theta}\right) = \frac{1}{8}(1 \cdot 1 \cdot 1) = \frac{1}{8}.$$

67.
$$\lim_{\theta \to 0} \frac{4\theta^2 + 3\theta \sin(\theta)}{\theta^2} = \lim_{\theta \to 0} \frac{\theta(4\theta + 3\sin(\theta))}{\theta^2} = \lim_{\theta \to 0} \frac{4\theta + 3\sin(\theta)}{\theta} = \lim_{\theta \to 0} \left(\frac{4\theta}{\theta} + \frac{3\sin(\theta)}{\theta}\right) =$$

$$= \lim_{\theta \to 0} \left(4 + \frac{3\sin(\theta)}{\theta}\right) = \lim_{\theta \to 0} 4 + \lim_{\theta \to 0} \frac{3\sin(\theta)}{\theta} = 4 + 3\lim_{\theta \to 0} \frac{\sin(\theta)}{\theta} = 4 + 3(1) = 4 + 3 = 7.$$

68.
$$\lim_{\theta \to 0} \frac{\sin[\cos(\theta)]}{\sec(\theta)} = \frac{\displaystyle\lim_{\theta \to 0} \sin[\cos(\theta)]}{\displaystyle\lim_{\theta \to 0} \sec(\theta)} = \frac{\sin\left(\displaystyle\lim_{\theta \to 0} \cos(\theta)\right)}{1} = \frac{\sin(\cos(0))}{1} = \sin(1).$$

69.
$$\lim_{\theta \to 0} \frac{\theta^2}{1 - \cos^2(2\theta)} = \lim_{\theta \to 0} \frac{\theta^2}{\sin^2(2\theta)} = \lim_{\theta \to 0} \frac{\theta}{\sin(2\theta)} \cdot \frac{\theta}{\sin(2\theta)} = \lim_{\theta \to 0} \frac{\theta}{\sin(2\theta)} \cdot \frac{2}{2} \cdot \frac{\theta}{\sin(2\theta)} \cdot \frac{2}{2} =$$

$$= \lim_{\theta \to 0} \frac{2\theta}{\sin(2\theta)} \cdot \frac{2\theta}{\sin(2\theta)} \cdot \frac{1}{4} = \frac{1}{4}\left(\lim_{\theta \to 0} \frac{2\theta}{\sin(2\theta)} \cdot \lim_{\theta \to 0} \frac{2\theta}{\sin(2\theta)}\right) \overset{\#1}{=} \frac{1}{4}(1 \cdot 1) = \frac{1}{4}.$$

Alternate solution:

$$\lim_{\theta\to 0}\frac{\theta^2}{1-\cos^2(2\theta)}=\lim_{\theta\to 0}\frac{\theta^2}{\sin^2(2\theta)}=\lim_{\theta\to 0}\frac{\theta}{\sin(2\theta)}\cdot\frac{\theta}{\sin(2\theta)}=$$

$$=\lim_{\theta\to 0}\frac{\theta}{2\sin(\theta)\cos(\theta)}\cdot\frac{\theta}{2\sin(\theta)\cos(\theta)}=\lim_{\theta\to 0}\frac{1}{4}\cdot\frac{\theta}{\sin(\theta)}\cdot\frac{\theta}{\sin(\theta)}\cdot\frac{1}{\cos(\theta)}\cdot\frac{1}{\cos(\theta)}=$$

$$=\frac{1}{4}\lim_{\theta\to 0}\frac{\theta}{\sin(\theta)}\cdot\lim_{\theta\to 0}\frac{\theta}{\sin(\theta)}\cdot\lim_{\theta\to 0}\frac{1}{\cos(\theta)}\cdot\lim_{\theta\to 0}\frac{1}{\cos(\theta)}\overset{\#1}{=}\frac{1}{4}\cdot 1\cdot 1\cdot\frac{\lim_{\theta\to 0}1}{\lim_{\theta\to 0}\cos(\theta)}\cdot\frac{\lim_{\theta\to 0}1}{\lim_{\theta\to 0}\cos(\theta)}=$$

$$=\frac{1}{4}\cdot\frac{1}{\cos(0)}\cdot\frac{1}{\cos(0)}=\frac{1}{4}\cdot\frac{1}{1}\cdot\frac{1}{1}=\frac{1}{4}.$$

70. $$\lim_{\theta\to 0}\frac{\sec(6\theta)\tan(3\theta)}{\theta}=\lim_{\theta\to 0}\frac{\tan(3\theta)}{\theta\cos(6\theta)}=\lim_{\theta\to 0}\frac{1}{\cos(6\theta)}\cdot\frac{\sin(3\theta)}{\theta\cos(3\theta)}=$$

$$=\lim_{\theta\to 0}\frac{1}{\cos(6\theta)}\cdot\frac{\sin(3\theta)}{\theta}\cdot\frac{1}{\cos(3\theta)}=\lim_{\theta\to 0}\frac{1}{\cos(6\theta)}\cdot\frac{\sin(3\theta)}{\theta}\cdot\frac{3}{3}\cdot\frac{1}{\cos(3\theta)}=$$

$$=3\lim_{\theta\to 0}\frac{1}{\cos(6\theta)}\cdot\frac{\sin(3\theta)}{3\theta}\cdot\frac{1}{\cos(3\theta)}=3\lim_{\theta\to 0}\frac{1}{\cos(6\theta)}\cdot\lim_{\theta\to 0}\frac{\sin(3\theta)}{3\theta}\cdot\lim_{\theta\to 0}\frac{1}{\cos(3\theta)}=$$

$$=3\cdot\frac{1}{\cos(0)}\cdot 1\cdot\frac{1}{\cos(0)}=3\cdot 1\cdot 1\cdot 1=3.$$

71. $$\lim_{\theta\to 0}\theta^2\cot^2(4\theta)=\lim_{\theta\to 0}\theta^2\cdot\frac{\cos^2(4\theta)}{\sin^2(4\theta)}=\lim_{\theta\to 0}\frac{\theta^2}{\sin^2(4\theta)}\cdot\cos^2(4\theta)=$$

$$=\lim_{\theta\to 0}\frac{\theta}{\sin(4\theta)}\cdot\frac{\theta}{\sin(4\theta)}\cdot\cos(4\theta)\cdot\cos(4\theta)=$$

$$=\lim_{\theta\to 0}\frac{\theta}{\sin(4\theta)}\cdot\frac{4}{4}\cdot\frac{\theta}{\sin(4\theta)}\cdot\frac{4}{4}\cdot\cos(4\theta)\cdot\cos(4\theta)=$$

$$=\frac{1}{16}\lim_{\theta\to 0}\frac{4\theta}{\sin(4\theta)}\cdot\frac{4\theta}{\sin(4\theta)}\cdot\cos(4\theta)\cdot\cos(4\theta)=$$

$$=\frac{1}{16}\lim_{\theta\to 0}\frac{4\theta}{\sin(4\theta)}\cdot\lim_{\theta\to 0}\frac{4\theta}{\sin(4\theta)}\cdot\lim_{\theta\to 0}\cos(4\theta)\cdot\lim_{\theta\to 0}\cos(4\theta)\overset{\#1}{=}\frac{1}{16}\cdot 1\cdot 1\cdot\cos(0)\cdot\cos(0)=$$

$$=\frac{1}{16}\cdot 1\cdot 1\cdot 1\cdot 1=\frac{1}{16}.$$

72. $$\lim_{\theta \to 0} \frac{\tan(\pi - \theta) - \theta}{\sin(\theta + \pi)} = \lim_{\theta \to 0} \frac{\frac{\sin(\pi - \theta)}{\cos(\pi - \theta)} - \theta}{\sin(\theta + \pi)} = \lim_{\theta \to 0} \frac{\left[\frac{\sin(\pi)\cos(\theta) - \cos(\pi)\sin(\theta)}{\cos(\pi)\cos(\theta) + \sin(\pi)\sin(\theta)}\right] - \theta}{\sin(\theta)\cos(\pi) + \cos(\theta)\sin(\pi)} =$$

$$= \lim_{\theta \to 0} \frac{\left[\frac{0 \cdot \cos(\theta) - (-1) \cdot \sin(\theta)}{(-1) \cdot \cos(\theta) + 0 \cdot \sin(\theta)}\right] - \theta}{\sin(\theta) \cdot (-1) + \cos(\theta) \cdot 0} = \lim_{\theta \to 0} \frac{\left[\frac{\sin(\theta)}{-\cos(\theta)}\right] - \theta}{-\sin(\theta)} = \lim_{\theta \to 0} \frac{\theta + \frac{\sin(\theta)}{\cos(\theta)}}{\frac{\sin(\theta)}{1}} =$$

$$= \lim_{\theta \to 0} \left[\frac{\theta}{\sin(\theta)} + \frac{1}{\cos(\theta)}\right] = \lim_{\theta \to 0} \frac{\theta}{\sin(\theta)} + \lim_{\theta \to 0} \frac{1}{\cos(\theta)} \overset{\#1}{=} 1 + \frac{1}{\cos(0)} = 1 + \frac{1}{1} = 1 + 1 = 2.$$

Let a and b be nonzero numbers.

73. $$\lim_{\theta \to 0} \frac{\cos(a\theta)\tan(a\theta)}{b\theta} = \lim_{\theta \to 0} \frac{\cos(a\theta)}{b\theta} \cdot \frac{\sin(a\theta)}{\cos(a\theta)} = \lim_{\theta \to 0} \frac{\sin(a\theta)}{b\theta} = \frac{1}{b}\lim_{\theta \to 0} \frac{\sin(a\theta)}{\theta} \cdot \frac{a}{a} =$$

$$= \frac{a}{b}\lim_{\theta \to 0} \frac{\sin(a\theta)}{a\theta} = \frac{a}{b} \cdot 1 = \frac{a}{b}.$$

Let a and b be nonzero numbers.

74. $$\lim_{\theta \to 0} \frac{\cos(a\theta)\tan(a\theta)}{\cos(b\theta)\tan(b\theta)} = \lim_{\theta \to 0} \frac{\cos(a\theta) \cdot \left[\frac{\sin(a\theta)}{\cos(a\theta)}\right]}{\cos(b\theta) \cdot \left[\frac{\sin(b\theta)}{\cos(b\theta)}\right]} = \lim_{\theta \to 0} \frac{\sin(a\theta)}{\sin(b\theta)} = \lim_{\theta \to 0} \frac{\sin(a\theta)}{\sin(b\theta)} \cdot \frac{ab\theta}{ab\theta} =$$

$$= \frac{a}{b}\lim_{\theta \to 0} \frac{\sin(a\theta)}{a\theta} \cdot \frac{b\theta}{\sin(b\theta)} = \frac{a}{b}\lim_{\theta \to 0} \frac{\sin(a\theta)}{a\theta} \cdot \lim_{\theta \to 0} \frac{b\theta}{\sin(b\theta)} \overset{\#1}{=} \frac{a}{b} \cdot 1 \cdot 1 = \frac{a}{b}.$$

75. $$\lim_{\theta \to 0} \frac{\sin^2(\theta) + 2\cos(\theta) - 2}{\cos^2(\theta) - \sin(\theta) - 1} = \lim_{\theta \to 0} \frac{1 - \cos^2(\theta) + 2\cos(\theta) - 2}{1 - \sin^2(\theta) - \sin(\theta) - 1} =$$

$$= \lim_{\theta \to 0} \frac{-[\cos^2(\theta) - 2\cos(\theta) + 1]}{-[\sin^2(\theta) + \sin(\theta)]} = \lim_{\theta \to 0} \frac{\cos^2(\theta) - 2\cos(\theta) + 1}{\sin^2(\theta) + \sin(\theta)} =$$

$$= \lim_{\theta \to 0} \frac{(\cos(\theta) - 1)(\cos(\theta) - 1)}{\sin(\theta)(\sin(\theta) + 1)} = \lim_{\theta \to 0} \frac{\cos(\theta) - 1}{1} \cdot \frac{\cos(\theta) - 1}{1} \cdot \frac{1}{\sin(\theta)} \cdot \frac{1}{\sin(\theta) + 1} =$$

$$= \lim_{\theta \to 0} \frac{\cos(\theta) - 1}{1} \cdot \frac{\theta}{\theta} \cdot \frac{\cos(\theta) - 1}{1} \cdot \frac{\theta}{\theta} \cdot \frac{1}{\sin(\theta)} \cdot \frac{1}{\sin(\theta) + 1} =$$

$$= \lim_{\theta \to 0} \frac{\cos(\theta) - 1}{\theta} \cdot \frac{\cos(\theta) - 1}{\theta} \cdot \frac{\theta}{\sin(\theta)} \cdot \frac{\theta}{\sin(\theta) + 1} =$$

$$= \lim_{\theta \to 0} \frac{\cos(\theta) - 1}{\theta} \cdot \lim_{\theta \to 0} \frac{\cos(\theta) - 1}{\theta} \cdot \lim_{\theta \to 0} \frac{\theta}{\sin(\theta)} \cdot \lim_{\theta \to 0} \frac{\theta}{\sin(\theta) + 1} =$$

$$= 0 \cdot 0 \cdot 1 \cdot \frac{\lim\limits_{\theta \to 0} \theta}{\lim\limits_{\theta \to 0} [\sin(\theta) + 1]} = 0 \cdot \frac{0}{0 + 1} = 0.$$

76. $$\lim_{\theta \to 0} \frac{\sin(2\theta) - \tan(2\theta)}{\theta^2} = \lim_{\theta \to 0} \frac{2\sin(\theta)\cos(\theta) - \frac{\sin(2\theta)}{\cos(2\theta)}}{\theta^2} =$$

$$\lim_{\theta \to 0} \left[2\sin(\theta)\cos(\theta) - \frac{2\sin(\theta)\cos(\theta)}{\cos^2(\theta) - \sin^2(\theta)} \right] \cdot \frac{1}{\theta^2} =$$

$$= \lim_{\theta \to 0} \frac{2\sin(\theta)\cos(\theta)[\cos^2(\theta) - \sin^2(\theta)] - 2\sin(\theta)\cos(\theta)}{\cos^2(\theta) - \sin^2(\theta)} \cdot \frac{1}{\theta^2} =$$

$$= \lim_{\theta \to 0} \frac{2\sin(\theta)\cos^3(\theta) - 2\sin^3(\theta)\cos(\theta) - 2\sin(\theta)\cos(\theta)}{[\cos^2(\theta) - \sin^2(\theta)]\theta^2} =$$

$$= \lim_{\theta \to 0} \frac{2\sin(\theta)\cos(\theta)[\cos^2(\theta) - \sin^2(\theta) - 1]}{[\cos^2(\theta) - \sin^2(\theta)]\theta^2} =$$

$$= \lim_{\theta \to 0} \frac{2\sin(\theta)\cos(\theta)[1 - \sin^2(\theta) - \sin^2(\theta) - 1]}{[\cos^2(\theta) - \sin^2(\theta)]\theta^2} = \lim_{\theta \to 0} \frac{2\sin(\theta)\cos(\theta)[-2\sin^2(\theta)]}{[\cos^2(\theta) - \sin^2(\theta)]\theta^2} =$$

$$= -4 \lim_{\theta \to 0} \frac{\sin^3(\theta)\cos(\theta)}{[\cos^2(\theta) - \sin^2(\theta)]\theta^2} = -4 \lim_{\theta \to 0} \frac{\sin(\theta)}{\theta} \cdot \frac{\sin(\theta)}{\theta} \cdot \frac{\sin(\theta)\cos(\theta)}{[\cos^2(\theta) - \sin^2(\theta)]} =$$

$$= -4 \lim_{\theta \to 0} \frac{\sin(\theta)}{\theta} \cdot \lim_{\theta \to 0} \frac{\sin(\theta)}{\theta} \cdot \lim_{\theta \to 0} \frac{\sin(\theta)\cos(\theta)}{[\cos^2(\theta) - \sin^2(\theta)]} =$$

$$= -4 \cdot 1 \cdot 1 \cdot \frac{\lim\limits_{\theta \to 0} \sin(\theta)\cos(\theta)}{\lim\limits_{\theta \to 0} [\cos^2(\theta) - \sin^2(\theta)]} = -4 \cdot \frac{\sin(0)\cos(0)}{\cos^2(0) - \sin^2(0)} = -4 \cdot \frac{0 \cdot 1}{1 - 0} = -4 \cdot \frac{0}{1} =$$

$$= 0.$$

77. $$\lim_{\theta \to 0} \frac{\sin(\theta) - 2\theta}{\theta} = \lim_{\theta \to 0} \frac{\sin(\theta)}{\theta} - \frac{2\theta}{\theta} = \lim_{\theta \to 0} \frac{\sin(\theta)}{\theta} - \lim_{\theta \to 0} 2 = 1 - 2 = -1.$$

78. $\displaystyle\lim_{\theta\to0}\frac{3-\csc(\theta)}{7-\cot(\theta)}=\lim_{\theta\to0}\frac{3-\dfrac{1}{\sin(\theta)}}{7-\dfrac{\cos(\theta)}{\sin(\theta)}}=\lim_{\theta\to0}\frac{\dfrac{3\sin(\theta)-1}{\sin(\theta)}}{\dfrac{7\sin(\theta)-\cos(\theta)}{\sin(\theta)}}=$

$\displaystyle=\lim_{\theta\to0}\frac{3\sin(\theta)-1}{\sin(\theta)}\cdot\frac{\sin(\theta)}{7\sin(\theta)-\cos(\theta)}=\lim_{\theta\to0}\frac{3\sin(\theta)-1}{7\sin(\theta)-\cos(\theta)}=\frac{\displaystyle\lim_{\theta\to0}[3\sin(\theta)-1]}{\displaystyle\lim_{\theta\to0}[\sin(\theta)-\cos(\theta)]}=$

$\displaystyle=\frac{3\sin(0)-1}{\sin(0)-\cos(0)}=\frac{3\cdot0-1}{0-1}=\frac{-1}{-1}=1.$

79. $\displaystyle\lim_{\theta\to0}\frac{\theta\cos(\theta)-\sin(\theta)}{\theta}=\lim_{\theta\to0}\frac{\theta\cos(\theta)}{\theta}-\frac{\sin(\theta)}{\theta}=\lim_{\theta\to0}\cos(\theta)-\lim_{\theta\to0}\frac{\sin(\theta)}{\theta}=\cos(0)-1=$

$=1-1=0.$

80. $\displaystyle\lim_{\theta\to0}\frac{\sin(2\theta)\tan(\theta)}{3\theta}=\frac{1}{3}\lim_{\theta\to0}\frac{\sin(2\theta)}{\theta}\cdot\frac{\tan(\theta)}{1}=\frac{1}{3}\lim_{\theta\to0}\frac{\sin(2\theta)}{\theta}\cdot\frac{2}{2}\cdot\frac{\tan(\theta)}{1}=$

$\displaystyle=\frac{2}{3}\lim_{\theta\to0}\frac{\sin(2\theta)}{2\theta}\cdot\frac{\tan(\theta)}{1}=\frac{2}{3}\cdot\lim_{\theta\to0}\frac{\sin(2\theta)}{2\theta}\cdot\lim_{\theta\to0}\tan(\theta)=\frac{2}{3}\cdot1\cdot\tan(0)=\frac{2}{3}\cdot0=0.$

81. $\displaystyle\lim_{\theta\to0}\frac{\sin(2\theta)+\tan(\theta)}{3\theta}=\frac{1}{3}\lim_{\theta\to0}\frac{\sin(2\theta)+\tan(\theta)}{\theta}=\frac{1}{3}\lim_{\theta\to0}\frac{\sin(2\theta)}{\theta}+\frac{\tan(\theta)}{\theta}=$

$\displaystyle=\frac{1}{3}\left[\lim_{\theta\to0}\frac{\sin(2\theta)}{\theta}+\lim_{\theta\to0}\frac{\tan(\theta)}{\theta}\right]\overset{\#29}{=}\frac{1}{3}\left[\lim_{\theta\to0}\frac{\sin(2\theta)}{\theta}\cdot\frac{2}{2}+1\right]=\frac{1}{3}\left[2\lim_{\theta\to0}\frac{\sin(2\theta)}{2\theta}+1\right]=$

$\displaystyle=\frac{1}{3}[2\cdot1+1]=\frac{1}{3}[2+1]=\frac{1}{3}\cdot3=1.$

82. $\displaystyle\lim_{\theta\to0}\frac{\tan(\theta)-\sin(\theta)}{\theta^2}=\lim_{\theta\to0}\frac{\dfrac{\sin(\theta)}{\cos(\theta)}-\dfrac{\sin(\theta)}{1}}{\theta^2}=\lim_{\theta\to0}\frac{\dfrac{\sin(\theta)-\sin(\theta)\cos(\theta)}{\cos(\theta)}}{\theta^2}=$

$\displaystyle=\lim_{\theta\to0}\frac{\sin(\theta)[1-\cos(\theta)]}{\theta^2\cos(\theta)}=\lim_{\theta\to0}\frac{\sin(\theta)}{\theta}\cdot\frac{1-\cos(\theta)}{\theta}\cdot\frac{1}{\cos(\theta)}=$

$\displaystyle=\lim_{\theta\to0}\frac{\sin(\theta)}{\theta}\cdot\lim_{\theta\to0}\frac{1-\cos(\theta)}{\theta}\cdot\lim_{\theta\to0}\frac{1}{\cos(\theta)}\overset{\#1}{=}1\cdot\lim_{\theta\to0}\frac{1-\cos(\theta)}{\theta}\cdot\lim_{\theta\to0}\frac{1}{\cos(\theta)}\overset{\#19}{=}0\cdot\frac{1}{\cos(0)}=$

$=0\cdot1=0.$

83. $\lim\limits_{\theta\to0}[\csc(\theta)-\cot(\theta)]=\lim\limits_{\theta\to0}\left[\dfrac{1}{\sin(\theta)}-\dfrac{\cos(\theta)}{\sin(\theta)}\right]=\lim\limits_{\theta\to0}\dfrac{1-\cos(\theta)}{\sin(\theta)}=$

$=\lim\limits_{\theta\to0}\dfrac{1-\cos(\theta)}{\sin(\theta)}\cdot\dfrac{1+\cos(\theta)}{1+\cos(\theta)}=\lim\limits_{\theta\to0}\dfrac{1-\cos^2(\theta)}{\sin(\theta)[1+\cos(\theta)]}=\lim\limits_{\theta\to0}\dfrac{\sin^2(\theta)}{\sin(\theta)[1+\cos(\theta)]}=$

$=\lim\limits_{\theta\to0}\dfrac{\sin(\theta)}{1+\cos(\theta)}=\dfrac{\lim\limits_{\theta\to0}\sin(\theta)}{\lim\limits_{\theta\to0}[1+\cos(\theta)]}=\dfrac{\sin(0)}{1+\cos(0)}=\dfrac{0}{1+1}=0.$

84. $\lim\limits_{\theta\to0}\left[\dfrac{1}{\theta^2}-\dfrac{1}{\theta^2\sec(\theta)}\right]=\lim\limits_{\theta\to0}\dfrac{\sec(\theta)-1}{\theta^2\sec(\theta)}=\lim\limits_{\theta\to0}\dfrac{\sec(\theta)-1}{\theta^2\sec(\theta)}\cdot\dfrac{\sec(\theta)+1}{\sec(\theta)+1}=$

$=\lim\limits_{\theta\to0}\dfrac{\sec^2(\theta)-1}{\theta^2\sec(\theta)\,[\sec(\theta)+1]}=\lim\limits_{\theta\to0}\dfrac{\tan^2(\theta)\cos(\theta)}{\theta^2[\sec(\theta)+1]}=\lim\limits_{\theta\to0}\dfrac{\tan(\theta)}{\theta}\cdot\dfrac{\tan(\theta)}{\theta}\cdot\dfrac{\cos(\theta)}{\sec(\theta)+1}=$

$=\lim\limits_{\theta\to0}\dfrac{\tan(\theta)}{\theta}\cdot\lim\limits_{\theta\to0}\dfrac{\tan(\theta)}{\theta}\cdot\lim\limits_{\theta\to0}\dfrac{\cos(\theta)}{\sec(\theta)+1}\overset{\#31}{=}1\cdot1\cdot\dfrac{\cos(0)}{\sec(0)+1}=\dfrac{1}{1+1}=\dfrac{1}{2}.$

85. $\lim\limits_{\theta\to0}\dfrac{\sin(\theta)}{\theta+\theta^2}=\lim\limits_{\theta\to0}\dfrac{\sin(\theta)}{\theta(1+\theta)}=\lim\limits_{\theta\to0}\dfrac{\sin(\theta)}{\theta}\cdot\dfrac{1}{1+\theta}=\lim\limits_{\theta\to0}\dfrac{\sin(\theta)}{\theta}\cdot\lim\limits_{\theta\to0}\dfrac{1}{1+\theta}=1\cdot\dfrac{1}{1+0}=$

$=1\cdot1=1.$

86. $\lim\limits_{\theta\to0}\dfrac{\sec(\theta)-1}{\theta^2}=\lim\limits_{\theta\to0}\dfrac{\sec(\theta)-1}{\theta^2}\cdot\dfrac{\sec(\theta)+1}{\sec(\theta)+1}=\lim\limits_{\theta\to0}\dfrac{\sec^2(\theta)-1}{\theta^2[\sec(\theta)+1]}=\lim\limits_{\theta\to0}\dfrac{\tan^2(\theta)}{\theta^2[\sec(\theta)+1]}=$

$=\lim\limits_{\theta\to0}\dfrac{\tan(\theta)}{\theta}\cdot\dfrac{\tan(\theta)}{\theta}\cdot\dfrac{1}{\sec(\theta)+1}=\lim\limits_{\theta\to0}\dfrac{\tan(\theta)}{\theta}\cdot\lim\limits_{\theta\to0}\dfrac{\tan(\theta)}{\theta}\cdot\lim\limits_{\theta\to0}\dfrac{1}{\sec(\theta)+1}\overset{\#31}{=}$

$\overset{\#31}{=}=1\cdot1\cdot\dfrac{1}{1+1}=\dfrac{1}{2}.$

87. $\lim\limits_{\theta\to0}\dfrac{\cos(2\theta)-\cos(\theta)}{\sin^2(\theta)}=\lim\limits_{\theta\to0}\dfrac{\cos^2(\theta)-\sin^2(\theta)-\cos(\theta)}{\sin^2(\theta)}=$

$=\lim\limits_{\theta\to0}\dfrac{\cos^2(\theta)-[1-\cos^2(\theta)]-\cos(\theta)}{\sin^2(\theta)}=\lim\limits_{\theta\to0}\dfrac{\cos^2(\theta)-1+\cos^2(\theta)-\cos(\theta)}{\sin^2(\theta)}=$

$=\lim\limits_{\theta\to0}\dfrac{2\cos^2(\theta)-\cos(\theta)-1}{\sin^2(\theta)}=\lim\limits_{\theta\to0}\dfrac{2\cos^2(\theta)-2\cos(\theta)+\cos(\theta)-1}{\sin^2(\theta)}=$

$=\lim\limits_{\theta\to0}\dfrac{2\cos(\theta)[\cos(\theta)-1]+\cos(\theta)-1}{\sin^2(\theta)}=\lim\limits_{\theta\to0}\dfrac{[\cos(\theta)-1][2\cos(\theta)+1]}{\sin^2(\theta)}=$

$$= \lim_{\theta \to 0} \frac{[\cos(\theta) - 1][2\cos(\theta) + 1]}{\sin^2(\theta)} = \lim_{\theta \to 0} \frac{[\cos(\theta) - 1][2\cos(\theta) + 1]}{\sin^2(\theta)} \cdot \frac{\cos(\theta) + 1}{\cos(\theta) + 1} =$$

$$= \lim_{\theta \to 0} \frac{[\cos^2(\theta) - 1][2\cos(\theta) + 1]}{\sin^2(\theta)[\cos(\theta) + 1]} = \lim_{\theta \to 0} \frac{-[1 - \cos^2(\theta)][2\cos(\theta) + 1]}{\sin^2(\theta)[\cos(\theta) + 1]} =$$

$$= \lim_{\theta \to 0} \frac{-\sin^2(\theta)[2\cos(\theta) + 1]}{\sin^2(\theta)[\cos(\theta) + 1]} = \lim_{\theta \to 0} \frac{-[2\cos(\theta) + 1]}{\cos(\theta) + 1} = \frac{\lim_{\theta \to 0} -[2\cos(\theta) + 1]}{\lim_{\theta \to 0} [\cos(\theta) + 1]} =$$

$$= \frac{-[2\cos(0) + 1]}{\cos(0) + 1} = \frac{-(2 \cdot 1 + 1)}{1 + 1} = \frac{-3}{2}.$$

88. $\lim_{\theta \to 0} \dfrac{\cos(\theta)}{\csc(\theta)} = \lim_{\theta \to 0} \cos(\theta)\sin(\theta) = \lim_{\theta \to 0} \cos(\theta) \cdot \lim_{\theta \to 0} \sin(\theta) = \cos(0) \cdot \sin(0) = 1 \cdot 0 = 0.$

89. $\lim_{\theta \to 0} \dfrac{\theta^3}{\csc(\theta) + 1} = \lim_{\theta \to 0} \dfrac{\theta^3}{\csc(\theta) + 1} \cdot \dfrac{\csc(\theta) - 1}{\csc(\theta) - 1} = \lim_{\theta \to 0} \dfrac{\theta^3[\csc(\theta) - 1]}{\csc^2(\theta) - 1} = \lim_{\theta \to 0} \dfrac{\theta^3[\csc(\theta) - 1]}{\cot^2(\theta)} =$

$$= \lim_{\theta \to 0} \theta^3 \left[\frac{1}{\sin(\theta)} - 1\right] \tan^2(\theta) = \lim_{\theta \to 0} \left[\frac{\theta^3}{\sin(\theta)} - \theta^3\right] \frac{\sin^2(\theta)}{\cos^2(\theta)} =$$

$$= \lim_{\theta \to 0} \left[\frac{\theta^3 \sin^2(\theta)}{\sin(\theta) \cos^2(\theta)} - \frac{\theta^3 \sin^2(\theta)}{\cos^2(\theta)}\right] = \lim_{\theta \to 0} \frac{\theta^3 \sin^2(\theta)}{\sin(\theta) \cos^2(\theta)} - \lim_{\theta \to 0} \frac{\theta^3 \sin^2(\theta)}{\cos^2(\theta)} =$$

$$= \lim_{\theta \to 0} \frac{\theta^3 \sin(\theta)}{\cos^2(\theta)} - \lim_{\theta \to 0} \frac{\theta^3 \sin^2(\theta)}{\cos^2(\theta)} = \frac{0^3 \sin(0)}{\cos^2(0)} - \frac{0^3 \sin^2(0)}{\cos^2(0)} = \frac{0 \cdot 0}{1} - \frac{0 \cdot 0}{1} = 0 - 0 = 0.$$

90. $\lim_{\theta \to 0} 2\theta^2 \sec^2(\theta) \cot^2(\theta) = 2 \lim_{\theta \to 0} \dfrac{\theta^2}{\cos^2(\theta)} \cdot \dfrac{\cos^2(\theta)}{\sin^2(\theta)} = 2 \lim_{\theta \to 0} \dfrac{\theta^2}{\sin^2(\theta)} \overset{\#1}{=} 2 \cdot 1 = 2.$

91. $\lim_{\theta \to 0} \dfrac{\cot^4(\theta) \tan(\theta) + \sin^2(\theta) - \csc(\theta) + \sec(\theta) =}{\theta^{-3}}$

$$\lim_{\theta \to 0} \theta^3[\cot^4(\theta) \tan(\theta) + \sin^2(\theta) - \csc(\theta) + \sec(\theta)] =$$

$$= \lim_{\theta \to 0} \theta^3 \left[\frac{\cos^4(\theta)}{\sin^4(\theta)} \cdot \frac{\sin(\theta)}{\cos(\theta)} + \sin^2(\theta) - \frac{1}{\sin(\theta)} + \frac{1}{\cos(\theta)}\right] =$$

$$= \lim_{\theta \to 0} \theta^3 \left[\frac{\cos^3(\theta)}{\sin^3(\theta)} + \sin^2(\theta) - \frac{1}{\sin(\theta)} + \frac{1}{\cos(\theta)}\right] =$$

$$= \lim_{\theta \to 0} \frac{\theta^3 \cos^3(\theta)}{\sin^3(\theta)} + \lim_{\theta \to 0} \theta^3 \sin^2(\theta) - \lim_{\theta \to 0} \frac{\theta^3}{\sin(\theta)} + \lim_{\theta \to 0} \frac{\theta^3}{\cos(\theta)} =$$

$$= \lim_{\theta \to 0} \frac{\theta^3}{\sin^3(\theta)} \cdot \lim_{\theta \to 0} \frac{\cos^3(\theta)}{1} + \lim_{\theta \to 0} \theta^3 \cdot \lim_{\theta \to 0} \sin^2(\theta) - \lim_{\theta \to 0} \frac{\theta}{\sin(\theta)} \cdot \lim_{\theta \to 0} \theta^2 + \lim_{\theta \to 0} \frac{\theta^3}{\cos(\theta)} =$$

$$= \lim_{\theta \to 0} \frac{\theta^3}{\sin^3(\theta)} \cdot \frac{\cos^3(0)}{1} + 0^3 \cdot \sin^2(0) - \lim_{\theta \to 0} \frac{\theta}{\sin(\theta)} \cdot 0^2 + \frac{0^3}{\cos(0)} \overset{\#1}{\cong} 1 \cdot \frac{1}{1} + 0 - 1 \cdot 0 + \frac{0}{1} =$$

$$= 1 + 0 - 0 + 0 = 1.$$

92. $$\lim_{\theta \to 0} \left[3\sec(\theta) - \frac{\theta^3 \csc^3(\theta)}{\cos^3(\theta)} + \theta^2 \csc(\theta) \right] = \lim_{\theta \to 0} \left[\frac{3}{\cos(\theta)} - \frac{\theta^3}{\sin^3(\theta) \cos^3(\theta)} + \frac{\theta^2}{\sin(\theta)} \right] =$$

$$= \lim_{\theta \to 0} \frac{3}{\cos(\theta)} - \lim_{\theta \to 0} \frac{\theta^3}{\sin^3(\theta) \cos^3(\theta)} + \lim_{\theta \to 0} \frac{\theta^2}{\sin(\theta)} =$$

$$= \lim_{\theta \to 0} \frac{3}{\cos(\theta)} - \lim_{\theta \to 0} \frac{\theta^3}{\sin^3(\theta)} \cdot \lim_{\theta \to 0} \frac{1}{\cos^3(\theta)} + \lim_{\theta \to 0} \frac{\theta^2}{\sin(\theta)} =$$

$$= \frac{3}{\cos(0)} - \lim_{\theta \to 0} \frac{\theta^3}{\sin^3(\theta)} \cdot \frac{1}{\cos^3(0)} + \lim_{\theta \to 0} \frac{\theta^2}{\sin(\theta)} \overset{\#1}{\cong} \frac{3}{1} - 1 \cdot \frac{1}{1} \cdot \frac{1}{1} + 1 = 3 - 1 + 1 = 3.$$

93. $$\lim_{\theta \to 0} \left[\cos(\theta) - \sin^3(\theta) \csc^2(\theta) - \tan(\theta) \right] = \lim_{\theta \to 0} \left[\cos(\theta) - \frac{\sin^3(\theta)}{\sin^2(\theta)} - \tan(\theta) \right] =$$

$$\lim_{\theta \to 0} \cos(\theta) - \lim_{\theta \to 0} \sin(\theta) - \lim_{\theta \to 0} \tan(\theta) = \cos(0) - \sin(0) - \tan(0) = 1 - 0 - 0 = 1.$$

94. $$\lim_{\theta \to 0} \left[8\theta^2 \csc^2(\theta) + \tan(\theta) \cos(\theta) \right] = \lim_{\theta \to 0} \left[\frac{8\theta^2}{\sin^2(\theta)} + \sin(\theta) \right] =$$

$$= 8 \lim_{\theta \to 0} \frac{\theta^2}{\sin^2(\theta)} + \lim_{\theta \to 0} \sin(\theta) \overset{\#1}{\cong} 8 \cdot 1 + \sin(0) = 8 + 0 = 8.$$

95. $$\lim_{\theta \to 0} 2\theta \cot(\theta) \sec(\theta) = 2 \lim_{\theta \to 0} \frac{\theta}{\sin(\theta)} \cdot \frac{\cos(\theta)}{1} \cdot \frac{1}{\cos(\theta)} \overset{\#1}{\cong} 2 \cdot 1 \cdot \frac{\cos(0)}{1} \cdot \frac{1}{\cos(0)} =$$

$$= 2 \cdot 1 \cdot \frac{1}{1} \cdot \frac{1}{1} = 2.$$

96. $\lim\limits_{\theta \to 0}\left[\dfrac{\cot(\theta)}{\csc(\theta)} + \sec(\theta)\right] = \lim\limits_{\theta \to 0}\left[\dfrac{\sin(\theta)}{1} \cdot \dfrac{\cos(\theta)}{\sin(\theta)} + \dfrac{1}{\cos(\theta)}\right] = \lim\limits_{\theta \to 0}\left[\cos(\theta) + \dfrac{1}{\cos(\theta)}\right] =$

$= 1 + 1 = 2.$

97. $\lim\limits_{\theta \to 0}\dfrac{\sin(2\theta)\cos^3(\theta)}{\sin(\theta)} = \lim\limits_{\theta \to 0}\dfrac{2\sin(\theta)\cos(\theta)\cos^3(\theta)}{\sin(\theta)} = 2\lim\limits_{\theta \to 0}\cos^4(\theta) = 2\cos^4(0) =$

$= 2 \cdot 1 = 2.$

98. $\lim\limits_{\theta \to 0}\left[\cos^2(\theta) - \sec(\theta)\sin(\theta)\right] = \lim\limits_{\theta \to 0}\left[\cos^2(\theta) - \dfrac{\sin(\theta)}{\cos(\theta)}\right] = \cos^2(0) - \dfrac{\sin(0)}{\cos(0)} = 1 - \dfrac{0}{1} = 1.$

99. $\lim\limits_{\theta \to 0}\dfrac{\cot^2(\theta) + 1}{\csc^2(\theta)} = \lim\limits_{\theta \to 0}\dfrac{\dfrac{\cos^2(\theta)}{\sin^2(\theta)} + 1}{\dfrac{1}{\sin^2(\theta)}} = \lim\limits_{\theta \to 0}\dfrac{\left(\dfrac{\cos^2(\theta) + \sin^2(\theta)}{\sin^2(\theta)}\right)}{\left(\dfrac{1}{\sin^2(\theta)}\right)} =$

$= \lim\limits_{\theta \to 0}\left[\cos^2(\theta) + \sin^2(\theta)\right] = \left[\cos^2(0) + \sin^2(0)\right] = 1 + 0 = 1.$

Alternate solution:

$\lim\limits_{\theta \to 0}\dfrac{\cot^2(\theta) + 1}{\csc^2(\theta)} = \lim\limits_{\theta \to 0}\dfrac{\csc^2(\theta)}{\csc^2(\theta)} = \lim\limits_{\theta \to 0} 1 = 1.$

100. $\lim\limits_{\theta \to 0}\dfrac{\csc^2(\theta)}{\cot^2(\theta) + 1} = \lim\limits_{\theta \to 0}\dfrac{\dfrac{1}{\sin^2(\theta)}}{\dfrac{\cos^2(\theta)}{\sin^2(\theta)} + 1} = \lim\limits_{\theta \to 0}\dfrac{\dfrac{1}{\sin^2(\theta)}}{\dfrac{\cos^2(\theta)}{\sin^2(\theta)} + \dfrac{\sin^2(\theta)}{\sin^2(\theta)}} =$

$= \lim\limits_{\theta \to 0}\dfrac{\left(\dfrac{1}{\sin^2(\theta)}\right)}{\left(\dfrac{\cos^2(\theta) + \sin^2(\theta)}{\sin^2(\theta)}\right)} = \lim\limits_{\theta \to 0}\dfrac{\left(\dfrac{1}{\sin^2(\theta)}\right)}{\left(\dfrac{1}{\sin^2(\theta)}\right)} = \lim\limits_{\theta \to 0} 1 = 1.$

Alternate solution:

$\lim\limits_{\theta \to 0}\dfrac{\csc^2(\theta)}{\cot^2(\theta) + 1} = \lim\limits_{\theta \to 0}\dfrac{\csc^2(\theta)}{\csc^2(\theta)} = \lim\limits_{\theta \to 0} 1 = 1.$

101. $\displaystyle\lim_{\theta \to 0} \frac{\sec^2(\theta)}{\tan^2(\theta) + 1} = \lim_{\theta \to 0} \frac{\dfrac{1}{\cos^2(\theta)}}{\dfrac{\sin^2(\theta)}{\cos^2(\theta)} + 1} = \lim_{\theta \to 0} \frac{\dfrac{1}{\cos^2(\theta)}}{\dfrac{\sin^2(\theta)}{\cos^2(\theta)} + \dfrac{\cos^2(\theta)}{\cos^2(\theta)}} =$

$\displaystyle = \lim_{\theta \to 0} \frac{\left(\dfrac{1}{\cos^2(\theta)}\right)}{\left(\dfrac{\sin^2(\theta) + \cos^2(\theta)}{\cos^2(\theta)}\right)} = \lim_{\theta \to 0} \frac{\left(\dfrac{1}{\cos^2(\theta)}\right)}{\left(\dfrac{1}{\cos^2(\theta)}\right)} = \frac{\left(\dfrac{1}{\cos^2(0)}\right)}{\left(\dfrac{1}{\cos^2(0)}\right)} = \frac{\left(\dfrac{1}{1}\right)}{\left(\dfrac{1}{1}\right)} = 1.$

Alternate solution:

$\displaystyle\lim_{\theta \to 0} \frac{\sec^2(\theta)}{\tan^2(\theta) + 1} = \lim_{\theta \to 0} \frac{\sec^2(\theta)}{\sec^2(\theta)} = \lim_{\theta \to 0} 1 = 1.$

www.ingramcontent.com/pod-product-compliance
Lightning Source LLC
Chambersburg PA
CBHW080621180526
45168CB00007B/3002